Harper and Bros.

Major General J. F. C. Fuller

ARMORED WARFARE

An Annotated Edition of

LECTURES ON F. S. R. III.

(Operations Between Mechanized Forces)

BY

MAJOR-GENERAL J. F. C. FULLER

C.B., C.B.E., D.S.O.

Foreword by

Lt. Col. S. L. A. Marshall, AUS.

GREENWOOD PRESS, PUBLISHERS
WESTPORT, CONNECTICUT

Library of Congress Cataloging in Publication Data

Fuller, J. F. C. (John Frederick Charles), 1878-1966.
Armored warfare.

Reprint. Originally published: 1st American ed.
Harrisburg, Pa. : Military Service Pub. Co., 1943.
1. Tank warfare. 2. Mechanization, Military.
3. Strategy. I. Fuller, J. F. C. (John Frederick
Charles), 1878-1966. Lectures on F.S.R. III (operations
between mechanized forces) II. Title.
UG446.5.F78 1983 358'.18 83-8247
ISBN 0-313-24067-1 (lib. bdg.)

Originally published in 1943 by the Military Service
Publishing Company, Harrisburg, Pennsylvania

Reprinted in 1983 by Greenwood Press
An imprint of Greenwood Publishing Group, Inc.
88 Post Road West, Westport, Connecticut 06881

Printed in the United States of America

10 9 8 7 6 5 4 3

Contents

Diagrams

Most of these diagrams are purely geometrical, their object being to express tactical ideas and not actual operations.

FOREWORD

Almost three years have passed since I first wrote to General Fuller urging that the times called for the publishing of a new edition of "FSR III" in which its theory could be profiled against the record of the new war. The conflict which he had foreseen so clearly was then in its early stages. The tactical and strategical concepts which he had committed to print in 1932 (dictating them to his stenographer while on a skiing trip in Switzerland) had been partially tested in battles and campaigns. The body of his work had been proved a nigh-faultless estimate, though certain of the collateral ideas, especially as regards the future of infantry and the importance of cities as defensive pivots against armored forces, warranted re-examination and continued study in the light of the war's lengthening experience.

I pressed it upon him as an obligation that he should amend his prophetic study of the nature of armored power and of the characteristics of warfare in our time. The greatest tank battles of the age were only then beginning in Africa. The most far-sighted military thinker of the Machine Age was then living in retirement in Surrey. It seemed a strangely ironic situation and the irony of it was not lessened by his reply to my letter: "You may imagine how far my thoughts are from such matters when I tell you that I am now writing a paper on Chief Pontiac's attack on the town of Detroit."

Yet at last the book is here and the publication of it is partial measure of the recognition newly come to this major prophet of war during his life-time. One reflects how easily it might have been otherwise had the man himself been inclined to brood over the past rather than to contend with the problems of the present. At war's beginning, he was the Forgotten Warrior. His political ideas had led him down a strange path. His military career was in eclipse. The most prophetic of his great writings were gathering dust because they had been written in military style to convince soldiers rather than in popular style to interest laymen, and it is notorious that soldiers do not read books. In fact, our own "Infantry Journal" received but failed to review "FSR III" at the time of its original publication; in fact, not more than half a dozen officers had even taken it from the Library at the Infantry School up to the day when the present war began. It had reposed there virtually undiscovered though it was undoubtedly the greatest treasure in the place. For nowhere else in the writings of men was the outline of the military future so clearly revealed and nowhere else had the tactics of modern battle been explored as brilliantly in conformity with war's enduring principles. One cannot weigh the pure gold contained in this slender volume without reflecting on what might have been, had the democracies ten years ago given it as much consideration as any one of the ten best sellers of that day.

Probably it is not remarkable at all that Fuller's ideas are having greater circulation today than ever before and that the book scorned in 1932 is at last receiving its just due. Almost too late, we have realized the magnitude of our error, and as a nation, we are intent on catching up with our military reading. To run and think at the same time is not conducive to the best thought, but we have no choice. A course has to be run all the way from our historical past into our technological present, if we are to survive this war with any greater understanding and firmer resolve not to relapse at once into the old slothful way of going. It is a distance far longer than many think and in our efforts to cover it, much time is wasted. One recalls the complaint of Napoleon that there were so many military books and it was necessary to read so many bad ones to find the few which were worthwhile. In the past three years there has been a veritable spate of resurrected volumes most of which sound no warning because it is much too late for that. Few of these books have any relation to our present situation. They have, for the most part, a post-mortem quality. In essence, and for our present purposes, they are not more than curiosities in the library of war. The idle will collect them as such. The more studious will reflect upon their contents and will find therein the lesson that closed minds are a greater danger to a nation than closed arsenals and gun pits.

Among these re-published works, "FSR III" stands out like Mars at perihelion. Now is not the time to assign it a place among the classic works on warfare though the day will surely come when it will be acknowledged such honor. Our need now is not dead books but live books which can be put to work by thinking soldiers. It is not enough to point to war's eternal principles when we are at grips with a powerful enemy; while the fight goes on we must think with ever increasing directness about the application of these principles to the battle which must be won. Abstractions will not save us, but fire power and getting there first with the most effective men and weapons. Herein are the unique values in General Fuller's book. Let me quote from one of his recent letters:

"So far as I can make out, though we are well into the fourth year of the war, we, over here, have in no way fixed on an armored warfare doctrine. It is important that we should, and that you should also, because without such there is nothing to bite on, to amend or to improve. FSR III, in spite of its antiquity, may still provide a datum point, something to start from and to criticize.

"It is really surprising that the old British Tank Corps of 1918 was far more an armored force than any existing today. It was all armored, and in 1919, had the war continued, all its auxiliary machines would have been armored also, such as bridge-laying, mine-sweeping, self propelled artillery and infantry transporters. Further still, it would have been all tracked, including its supply vehicles, of which we were in the process of building 10,000. To all intents and purposes the wheel would have been eliminated. Nothing like such an organization exists today in any army. What soldiers and designers have done is to concentrate on fighting machines, and so far to leave all others out of the organic picture."

In 1932, when these lectures were first published, the opinion all but universally held was that it would not be possible without years of experimental work to decide on the organization and tactics of tanks. Thus, for example, a year later, General Douglas MacArthur, then Chief of Staff, United States Army, wrote in a Foreword to a book entitled—"The Fighting Tanks Since 1916": "We cannot today state categorically either their potential capacities or their ultimate limitations. It is a natural consequence of this situation that the doctrines for their employment are subject to continuous and at times fundamental changes."

This, Fuller held to be incorrect, reasoning rather that though the ultimate limitations of tanks were unknown, their potential capacities were either known or predictable. It was his view that though changes would progressively arise, radical ones were unlikely.

Working on the theory he had always worked on when considering tactics—namely, that its elements are movement, protection and offensive action—he elaborated these lectures. And though he had nothing to found them on except these elements, and such experience as he had gained during the last war, it suffices to say that if he were called upon to wholly revise the book today in view of the experience gained in the last decade, the required changes would be neither multitudinous nor revolutionary. Except for a few verbal amendments, therefore, the lectures are as they were originally written, save that a number of footnotes have been added to bring them up to date or to show when and where their doctrines have been proved right or wrong during the present war.

It seems to one that this is the proper principle though it is all too seldom used in instruction. For by eschewing revision and the temptation to "bring things up to date" in writings which are based with fidelity upon the principles of war, there is the better chance to impress the student with the enduring values of principles instead of causing him to reflect on the dubious blessings of hindsight. The few in America who were familiar with "FSR III" during these recent years and who used it as a datum point need not be told of its worth as a working document on war.

Any who became saturated with its ideas indeed found it possible to foresee what would happen in the battle for the USSR and to forecast accurately the nature of the warfare in North Africa. There has been scarcely a development in armored warfare or in defense against tanks which was not envisaged by Fuller. In this vital respect, the document is unique. There have been other noted books on armored combat, but none which viewed the subject "in the round," as well as in detail, with such crystal clarity.

Again, however, it is to be noted that it is Fuller's regard for the far horizons of war which invariably makes his foregrounds comprehensible. His great powers as a critic reside equally in his vast knowledge of the history of war and his grasp of its practical difficulties, for without the latter no mere acquaintance with principles will suffice. As he himself has said: "No book can teach you to ride a bicycle; you have got to get on to the machine and you have got to fall from it many times before you can begin to think of becoming a champion." "FSR III," therefore, is not to be regarded as the writing of a theorist but as the digested wisdom of a field soldier who absorbed all that he saw in the first World War, screened the lessons from his first-hand experience and continued to think boldly about the future.

Though originally written for the British Army and around its military Bible—the Field Service Regulations—the lectures were more studied in Europe than in the English-speaking democracies. There were editions published for the information of the Red Army and the Spanish Army, and the Czechs thought so highly of "FSR III" that it became the standard work by which mechanical warfare was taught at the Czech Staff College. Of its influence in the German Army, a high Czech officer wrote: "The Germans regard it as their Bible, and the entire attack in France and Flanders was based on Fuller's teaching." In part corroboration of this, Brigadier T. Denis Daly, who in 1939 was British military attache in Berlin, wrote in "The Army Quarterly" of August 1942: "Another . . . was General Guderian who told me at length of the writings of Major General Fuller." Heinz Guderian was then Germany's foremost tank expert.

ARMORED WARFARE

A high placed Russian officer tells me that three months before the German attack on the USSR, Marshal Timoshenko ordered that "FSR III" be made a "table book" for the Red Army—to be kept at every officer's elbow and used for constant reference. He said that Clausewitz' "On War" and Douhet's "Command of the Air" were the only other books in this category.

When Fuller set out to write the lectures, as he himself tells the story, he had the following general picture in mind: All organized armies were in a state of transition, somewhat unwillingly, passing from a muscle-footing to a motor-footing, as during the nineteenth century navies had passed from a sailing-footing to a steaming footing. The further assumption was that though the muscle-moved forces were likely to continue for a long time, the motor-moved would assume two general forms—the armored and the unarmored. In tactics the first would be "regular" and the second "irregular"; hence Fuller called the second "motor guerillas." By this he did not mean "franc tireurs," but soldiers who, though they belonged to properly constituted military units or formations, would fight in an iregular way—that is, more independently than the armored troops, and much as the Germans used their motor cyclists during their invasion of France. At the time General Fuller was somewhat influenced by General von Seeckt's proposal to reinforce the small, long service and highly trained German Reichswehr by a militia embracing the mass of the manpower of the country, but with this difference: Whilst the professional army Fuller looked for was to be mechanized—that is, armored—the militia was to be motorized and trained, anyhow in part, on guerrilla lines.

Again, though at the time it was held that assault tanks should be heavily armored, Fuller over-estimated the protective power of speed and in consequence under-estimated the likelihood of a general thickening of armor, and, in consequence, a rise in the caliber of tank and anti-tank artillery. Though he has never rated the light tank—such as was then being introduced—highly, it remains his opinion that the types laid down in these lectures are those required, and more especially the long distance and rapidly moving reconnaissance machine.

Though, since these lectures were written, armor has increased in thickness and guns have increased in caliber, speed and range of action are much as they were eleven years ago. Therefore, no fundamental tactical change has developed, and this being so, so long as the reader is acquainted with the administrative and organizational changes which mark these eleven years, he will still find these lectures, old as they are, of tactical value. In any case they do show that it is possible to develop sound tactics from all but purely theoretical data, and that, therefore, there is no military excuse why, in 1939, the democracies were so utterly unprepared to wage a war of machines. What today is wanted is a new tank bible, even though it might throw into obsolescence certain of the tactical arrangements herein described. Those who read these lectures will be best qualified to say whether J. F. C. Fuller is the man to write it. I feel certain that they will have some added satisfaction also from the information that the diagrams appearing in this book were all planned by Fuller.

The book is a challenge to the soldier to become thoughtful. It is an invitation to him to keep his glance directed toward the far horizons of war even though he labors now at a desk job in some unexalted post. Beyond our obligation as soldiers, each of us has his duty and obligation as a citizen. This, hard experience has taught us—or should have done so—that within a democracy, the military problems cannot be solved by the inspired thinking of a general. Every good idea starts with a minority of one, but it must pass from the one to the many before we can find the road to salvation. In terms of the military future, that means millions of people thinking more deeply about the safety of the nation and about the nature of military power in the modern world than they have ever thought in their lives.

We may each agree with the words of the late Justice Oliver Wendell Holmes: "In the midst of doubt, in the collapse of creeds, there is one thing I do not doubt, that no man who lives in the same world with most of us can doubt, and that is that the faith is true and adorable which leads a soldier to throw away his life in obedience to a blindly accepted duty, in a cause which he little understands, in a plan of campaign of which he has no notion, under tactics of which he does not see the use."

But it would be a truer faith if he clearly saw his duty, if he rejoiced in his understanding of the national cause and if he knew the why of tactics. A faith may stand or fall according to the willingness of men to die for it, but it acquires nobility in the degree that those who are most ready to die are likewise most zealous to think in behalf of it.

S. L. A. MARSHALL

Lieut. Colonel
Army of the United States.

July, 1943

PREFACE TO THE FIRST EDITION

THIS book, I believe, is the first of its kind—that is to say, the first complete manual yet written on operations between mechanized armies. It is true that in our army there does exist a confidential text book on armoured forces; but it is in no sense a complete manual, and being confidential it defeats the aim of all such works—namely, publicity. What is wanted to-day is a public and not a secret book, a book which may be obtained anywhere, read anywhere and left anywhere, and not a book coupled with lock and key, for such books remain under lock and key and are little read except by the curious.

For many years now I have urged that general tactical training for war demands two books, one dealing with present-day warfare, and the other with future warfare. I have suggested this time and again, because we are living in a transitional military age, and were war to break out to-morrow, unless we have carefully thought out future possibilities, however full our knowledge of actualities may be, we shall be most indifferently prepared to wage it.

The first of these books is represented by "Field Service Regulations, Vol II.," the second should be represented by "Field Service Regulations, Vol. III.," hence the title of these lectures; lectures written on a book which does not exist, but which should exist, and one day no doubt will exist.

Having no official guide to follow, the system I have adopted is to take my "Lectures on F. S. R. II.," and so far as it is possible "motorize" and "mechanize" each lecture. To have sat down and written a completely new manual would, in many ways, have been an easier task; but it would not have enabled the student so easily to relate the future to the present. In a new manual the reasoning would have been far more logical, and though the present volume is little more than a series of notes, it nevertheless does enable the reader to tackle the whole subject of war in a logical manner, namely, first to study "F. S. R. II.," the official textbook, next to study my "Lectures on F. S. R. II." which explain and comment on it, and lastly to study this present book which projects this explanation and comment into the future.

That this process of study will lead the student to an exact appreciation of the future I am the first to deny. I am not so presumptuous as to suppose that my views are correct, far from it; for I realize as fully as anyone that they must remain in solution until they are proved right, or wrong, by experience. But what I hold fast to is this: They will, I think, cast doubt upon many of our present certainties, and so establish new problems which cannot be shelved, but must be solved one way or the other; that is to say, they must be proved right or wrong, or what is more probable—inadequate, that is susceptible to improvement.

The method I have adopted in writing this book—namely, in hinging it upon "F. S. R. II."—has led to many repetitions in the book itself. Though on an old and well established subject this would be a defect, I do not consider it so here. Repetition is the foundation of instruction, and the repetitions which occur will anyhow accentuate the importance of certain points.

A correcter criticism is that I have dealt inadequately with the power of aircraft, by many considered **the** arm of the future. This is true, but my excuse is that I have been working within the framework of "F. S. R. II." Aircraft, a totally motorized arm, I believe to have an enormous future, a future which I hope some day to deal with in another book—a true manual on motorized and mechanized warfare.

Another point is, that the suggestions I have made may prove so costly as to be unpractical. My answer to this is that I have not attempted in this book to recreate the army, but instead to jog the minds of thinking soldiers. I have attempted to show what may happen, not what will happen. Pure science does not think of cost, though applied science may have to. What we want to-day is not so much to remodel the army as to remodel our ideas. As it is common-sense to cut your coat according to your cloth, so also it is common-sense to have some idea in your head of what kind of coat you want before you buy your cloth.

PREFACE

This is not so much a book of conclusions, as a book of ideas, and if by means of it some of the younger brains in our army are rendered more plastic to change, then its object will be fulfilled; for it is the rising generation of soldiers which really matters.

Should we be called upon, some time during the next twenty-five years, once again to take part in an European war, it is men of this generation who will fight our battles, and irrespective of the arms they may command and the weapons their men have been trained to use, unless their minds are as flexible as high grade steel and as impressionable as wax, they will fail as certainly as the generals of 1914 failed within a few days of the World War being declared.

This plasticity of mind cannot be cultivated during war, except by an occasional genius; the generality of soldiers simply cannot change if they are dogma-ridden. The only way to prevent this ossification of mind is to accept nothing as fixed, to realize that the circumstances of war are everchanging, and that, consequently, organization, administration, strategy and tactics must change also; and if during peace time we cannot change them in fact, we can nevertheless change them in theory, and so be mentally prepared when circumstances require that changes should be made.

Adherence to dogmas has destroyed more armies and lost more battles and lives than any other cause in war. No man of fixed opinions can make a good general; therefore, if this series of lectures, however hypothetical many of its contentions may seem, succeeds in unfixing dogmas, then certainly it will not have been written in vain.

J. F. C. F.

February, 1932.

LECTURE I

CHAPTER I

Armed Forces, Their Command and the Principles of War

I. Introduction

To-day every organized army is faced by the greatest revolution that has ever taken place in the history of land warfare, a revolution which will parallel and perhaps exceed that accomplished by steam-power as applied to warfare at sea. It is true that steam-power led to a vast increase in the size of armies, and in the range and destructive effect of their weapons, as well as greatly extended their administrative and strategical powers; but it did not radically change their organization, for in spite of weapon improvement their tactics remained much the same.

It was not until the advent of the internal combustion engine that tactics began to assume strange and unexpected forms. First, the invention of aircraft introduced a totally new arm; secondly, the motor vehicle vastly extended the administrative power of armies and directly led to the reintroduction of armour; and thirdly, lethal, vesicant and lachrymatory gases and chemicals were placed on a definite weapon footing. As during the fifteenth and sixteenth centuries the changes in military organization and equipment were due to gunpowder, and those during the nineteenth century to steam-power and chemical science, so in the present century the radical changes must be sought in gasoline and electrical power, which, coupled with high explosives, steam-power and chemistry, cannot fail so completely to change the entire aspect of warfare, as to establish a new military dispensation.

It cannot be doubted, therefore, that to-day we are faced by so rapid a development, or evolution, in administration, strategy and tactics, and through these in organization, command and discipline, that this development constitutes a revolution which renders our existing art of war obsolete, so obsolete that unless we can grasp what it portends, to rely on it in another war is likely to prove a greater danger than to enter it totally ignorant of military values. The reason for this is that an open mind will readily adapt itself to circumstances, whereas one crowded with obsolete knowledge must first be disencumbered before it can adapt itself to its surroundings.

In the course of these lectures I shall attempt to outline the new theory of war which the internal combustion engine is creating, a theory founded on a new degree of movement.

Formerly, and until about a hundred years ago, all movement was muscular, and except for troop and supply movements by rail, it remained so until the First World War, when the truck and the motor car at once began to modify strategy and tactics. The first vastly increased the radius of supply by linking the railheads to the battlefields, and so rendered possible not only the great artillery engagements of that period, but enabled field fortifications to take on far more powerful forms than had ever been known before. The second largely extended means of intercommunication and personal inspection, for though generals and their staffs rode into the war on horse-back, immediately contact was gained with the enemy their horses were exchanged for motor cars, without which personal touch would have been rendered impossible.

The natural sequent of the motor car was the armoured car, and its immediate descendant the tracked armor car, or tank. The first of these two arms influenced strategy through extending reconnaissance, and the second radically influenced tactics for it cut out the bullet, which was the pivotal weapon of the older form of war. Whilst hitherto the bullet on the defense had proved itself to be, beyond all question more powerful than the bullet in the attack, through armour its power was reduced almost to zero, with the result that the only weapon the tank had to fear was the gun, which so long as it remains unarmoured, must, on account of the bullet, prove a frail protection to infantry attacked by tanks, as well as an indifferent anti-tank weapon.

Probably the greatest potential change which the First World War revealed was the power of aircraft. Strategically, the airplane opened up a totally new field of reconnaissance, tactically it not only radically modified artillery tactics, but, by being able to circumvent land forces, civil and military targets in rear of them were open to attack. It added a new dimension to the art of war, so full of possibilities that some think it may eventually grow so powerful as to render armies and navies useless. Even if this should not prove to be the case, it is a certainty that it will greatly modify them.

2. The Nature of War

As the weapons of war change, so does the character of war change, and though this is an undoubted fact, tactically it must not be overlooked that weapons change because civilization changes; they do not change on their own account. To-day wars arise out of economic causes, because our present civilization is essentially an economic one, its master being the machine in one form or another. As the present age is largely a mechanical one, so will the wars of this age take on a similar complexion, because military organization follows civil organization. When Europe was all but roadless, the horse was the civil and military means of movement, consequently cavalry abounded. When it became better roaded and fully agriculturalized, automatically infantry became the principal arm. So to-day, because manufacture is rapidly replacing agriculture as the main occupation of civilized nations, it may be taken as a certainty that military organization will follow suit, armies basing themselves more and more on the machine, the present expression of civil motive power.

What does this mean? It means that as until recently the most sturdy and reliable soldiers were drawn from the agricultural population, if not to-day, anyhow in the near future, all civilians connected during peace time with machines, and more particularly those which can immediately be used in war—motor cars, trucks, buses, tractors and civil airplanes—will form the main recruiting ground of armies.

What does this mean from the tactical point of view? As any man who has driven any kind of motor vehicle can rapidly be put through a modified course of musketry, or machine gun training, it means that in the future we are likely to be faced by two forms of field warfare: That carried out by highly organized armies, and that carried out by hastily organized guerilla forces.

[Throughout, the reader should consider the motor guerilla to be a motorized infantryman trained for independent action.]

In the South African War of 1899-1902, because every Boer was a horseman every Boer was a potential mounted infantryman; similarly to-day, every driver is a potential motorized guerilla. This being so, it is unlikely that, in another war, full use will not be made of such irregular forces; in fact the probability is that they will abound as they did during the eighteenth century, Croats, Pandours and Tyrolese riflemen being replaced by truck, bus and motor car drivers. How to make use of such forces, offensively and protectively, is one of the new and by no means least important problems of war which now faces us.

As industry is the base of mechanization, it logically follows that in the future only industrialized countries will be able to wage organized warfare with success. When war depended on horse-flesh, as it did during the Middle Ages, a country which possessed few horses stood a poor chance against a country possessing a plentiful supply. Similarly in those distant days, a country which could produce armour was all powerful when compared to a country which could not. So also to-day, a country which possesses few industries and manufactures few mechanical vehicles will be virtually impotent to resist invasion.

3. Armed Forces

If I am right in my assumption that industry is now the controlling factor in war, it follows that the whole of our present ideas on armed forces will have to be modified and recast. During what I will call "the agricultural period of war," power was sought through numbers, size being limited by facility of supply. Before the advent of the locomotive and the railway, armies remained comparatively small, because supply and evacuation depended upon road transport. Once steam-power came to the assistance of armies, their size grew so enormously that, in order to control their movements, strategy and tactics—governed as they are by supply—became exceedingly methodical, and organization following suit. Whereas in the eighteenth century barrack square drill was often the deciding tactical factor, in the later part of the nineteenth century, in order to facilitate control, individual initiative was largely replaced by doctrine and dogma. All commanders and leaders were drilled to think alike, and frequently without regard to actual or probable circumstances. The result was that any unexpected event, or miscalculation, was apt to throw the masses out of gear, as happened within a fortnight after the First World War opened.

The motorization and mechanization of armies reverse the whole of this process of organizing fighting and thinking. First, as I have shown, guerilla warfare, the most primitive of all forms of war, is likely to be revived, and as it obviously demands a high order of initiative to combat it, it will force this essential quality upon the commanders of organized forces. Secondly, the expense of raising mechanized armies will limit their size. Whereas in the First World War millions of infantry were mobilized, in the next war it is highly improbable that any nation will be able to put into the field more than a few thousand tanks.

[With the possible exception of Russia, this was true in 1939. It would appear that the Germans had about 5000 modern tanks; the Poles a few hundred; the Italians about 1200 and the French and British some 4000, of which the greater part was obsolete. On the outbreak of the war about three-quarters of all machines were light tanks.]

And as these machines can move far more rapidly than infantry in order to be in a position to seize the initiative on the outbreak of war, mechanized forces will have to be kept fully mobilized during peace time. This leads us to the third conclusion—namely, for fighting purposes highly trained professional armies will replace the present-day short service conscript masses. This does not necessarily mean that conscription will disappear, but that the conscripted man will become what may be called a soldier of the "second degree"—that is, a second grade fighter; one who will occupy, organize and hold the areas conquered by the mechanized forces. From this it may be predicted that, in the main, these soldiers will be a combination of pioneer and gendarme.

[See Lecture II, Section 10, Page 18.]

4. The Employment of the Armed Forces

Thus far we have arrived at three main categories of soldiers, not cavalry, artillery and infantry, but motorized guerillas, a mechanized first line and an unarmoured second line, part of which will be moved in motor vehicles and part will continue to march. We have here a curious picture, a picture which resembles the organization of the Macedonian army under Alexander the Great. The motorized guerillas may be compared to his light cavalry; the mechanized, to his heavy armoured cavalry and his phalanx of armoured foot; the second line of his auxiliary units, light troops and camp followers.

Here I will make a slight digression: Should the student wish to study the possibilities of future warfare, I can recommend him no better periods than those of Classical warfare, Medieval warfare and eighteenth century warfare, in the last of which free bands, or guerillas, abounded. Conversely, I can think of no worse period than that most frequently studied—the period which began with Spicheren in 1870 and ended at the battle of Cambrai on November 20, 1917.

At Cambrai what was the predominant value of the tank? It was its moral effect. It showed clearly that terror and not destruction was the true aim and end of armed forces. That is to say, to attack the nerves of an army, and through its nerves the will of its commander, is more profitable than battering to pieces the bodies of its men. The airplane also proved this, and even more dramatically; for not only could it attack the will of an army by avoiding its body, but also it could attack the political will and the national will behind the army. The air force constitutes, therefore, a fourth category of soldiers, which must never be divorced from the other three categories.

From these considerations—the power of aircraft to strike at the civil will, the power of mechanized forces to strike at the military will, and the power of motorized guerillas to broadcast dismay and confusion—we may predict that the power to effect physical destruction, which reached its zenith during the First World War, will gradualy and increasingly be replaced by attempts to demoralize the will of the enemy in its several forms, and so not only disorganize his armies but unnerve his people.

[We now know that the war on nerves was one of the main items in German war theory. It succeeded beyond measure in Norway, Holland, Belgium, France and in Great Britain immediately following the French collapse; but it failed in Russia. Therefore this prediction may be considered an over-statement.]

5. The Command of the Military Forces

It will be realized at once that all these many changes must have a profound influence upon generalship. At one time and not so far distant, as the Crimean War, the Indian Mutiny and the American Civil War distinctly show, the point of danger was the point of honour, and to this point a general instinctively went. Since 1870, largely on account of the ever-increasing size of armies and the complexity of their organization, generals have been forced to abdicate this honourable position and became administrators rather than commanders. Complexity rapidly increased the staffs, and these in their turn absorbed the little that remained in generalship, until in the First World War we find the old and rightly deprecated system of command by councils of war re-established under new and thinly disguised names, such as conferences and committees of experts.

During the First World War this system of command failed to produce a single general of the first order, not that such men did not exist, but that the circumstances in which they were compelled to work cramped their initiative. In a war of machines, aircraft moving at 200 miles an hour, motor vehicles at 40 and tanks at 20, such a system of command becomes insane. In the First World War it worked at great cost; in a mechanized war it simply cannot work at all. Therefore, once again we must get back to past history, to the time when generals really commanded and moralized their troops by sharing their dangers with them. This should not be so difficult as may at first appear; for, as mechanization proceeds, armies will grow smaller and not larger, and a more complete division will take place between what I have called soldiers of the first and second degrees, those who do the fighting and those who do the occupying. The latter will still mainly require an administrator and a highly organized administrative staff to control them, the former will require a general of high initiative with a small operational staff. The general will be **with** his fighting troops, he will be **in** the battle and not outside of it

[This has been proved correct as regards armoured formations, more particularly the German. In consequence they have lost many general officers.]

6. The Commander's Plan

As initiative, far more so than method, will prove to be the secret of success in all operations other than siege warfare, plans will have to be exceedingly simple and flexible. Much will have to be left to the initiative of subordinate commanders, consequently the leading idea of an operation must not only be known to all, but alternative movements must also be considered. Thus, for example: The leading idea may be to turn the enemy's right which is known to be at **a,** but unless the enemy is definitely pinned down at **a,** should his forces be motorized, or mechanized in an hour or two he may move them to **b,** and so render the plan of turning his right abortive. To prevent such a disruption of a plan, three things are apparent: First, if the enemy can be fixed at **a** he cannot move to **b;** secondly, if he cannot be fixed, then he must be kept under observation, so that the moment he moves his movement may be signalled back; thirdly, unless the plan includes alternative courses of action, so that direction may be changed **as the enemy moves,** much time will be lost, and every five minutes wasted may mean a mile or two on the debit side of the plan's account.

What is presented to us here? A picture closely resembling naval warfare. The initial plan is in the main an idea expressed in the simplest terms, and in accordance with it a temporary distribution is decided upon. The idea is then equated with the possible courses the enemy may adopt, let us suppose **a, b, c, d;**

for each of these courses a distribution of force is worked out, which I will call **w, x, y, z;** each subordinate commander being acquainted with the general idea as well as these alternative courses of action. Then, as the enemy moves, so will one of these courses be signalled to all units, who will act accordingly. There will then be no halt in the advance, or the attack, in a pursuit, or in a retirement; therefore no time will be lost. In brief, to economize time in action will become the soul of every plan.

[It would appear from the reports on the various Libyan campaigns that though General Rommel's object throughout was a fixed one—namely, the occupation of Egypt, each of his battle plans has been based on a series of alternative moves, the one finally decided on being dictated by the circumstances of the moment.]

In operations which for success, or economy of force, depend upon rapidity of movement, our present system of written and somewhat detailed operation orders will have to be replaced by some such system as I have just outlined. As frequently a situation may be such that all courses which the enemy may adopt cannot be foreseen, or if a certain course he may take is likely to upset the general idea altogether, I consider that it will be wise to have what I will call a "rally alternative". That is to say, should the enemy move in a direction which is likely to render all the alternatives laid down abortive, then, instead of making the best of a bad job, which with motorized and mechanized arms is very liable to lead to confusion and loss of control, the operations will be called off, the troops engaged at once assuming a defensive order.

[I am of opinion that had some such arrangement as this been agreed upon before General Cunningham attacked General Rommel on November 18, 1941, the confusion which appears to have characterized the first few days of fighting would have been considerably reduced. In former days the same applied to the handling of cavalry—e.g.: Cromwell at Marston Moor and Naseby.]

In brief, the plan will aim at developing the highest possible initiative without loss of control, command being maintained by alternative courses of action which are signalled to subordinate commanders by code letters. Liberty and responsibility are the abutments upon which the plan must rest, rather than method and obedience which hold good for slow moving infantry and artillery masses.

7. The Principles of War

I have now said enough about the probable changes in military organization and command to show how revolutionary they are likely to be, and how at variance they are with our present-day training. It would almost seem that the whole of it will have to be recast: is there then nothing permanent which we can hold on to? Fortunately, "yes," the principles of war; and directly it is realized that these principles form the foundation of mechanized warfare, just as they do of muscular warfare, it will be seen that revolution is really evolution. What we are faced with is not a new type of war, a war totally unrelated to the present type, but a new form of war, a form arising out of the internal combustion engine which has greatly accelerated movement and enhanced carrying power. There is nothing absolutely new under the sun, and as I have already pointed out, a study of history will reveal to the student many of the phases of war which will once again be passed through in almost identical forms. A little study and reflection will lead us to realize that all past strategy and tactics has been governed, consciously or unconsciously, as it may be, by the principles of war, and we may deduce from this that this government will continue. Economy of force, concentration, surprise, security, offensive action, movement and cooperation hold good whether an army is composed of foot-soldiers, horse-soldiers, or machine-soldiers. After all, it is surely obvious that motorization and mechanization are only changes in the conditions of war, in the tools which the general uses and not in the principles of his art. All this is so obvious that I will ask my listeners to accept the principles of war as they stand, and without modification, and further to look upon them as the sheet-anchor of their studies, whether they are examining past warfare, present warfare, or future warfare.

8. The Application of the Principles of War

The present revolution lies in the application of the principles of war to changing conditions. Of these conditions the two main ones are enhanced movement and protection—that is, speed and armour. Speed has increased from a horse galloping at 20 miles the hour to an airplane flying at 200, and a man walking at 4 miles the hour to a motor vehicle moving at 40, or a tank at 20; and armour has completely defeated the rifle bullet whether ordinary or armour-piercing. Though innumerable other changes in the conditions of war are taking place, these two are sufficient to show how radical this revolution is.

As regards the application of the principles of war, this will depend largely upon the organization and equipment of the enemy.

Three choices face us: (1) When one side is not mechanized, (2) when both are partially mechanized, and (3) when both are fully mechanized.

Against a non-mechanized enemy and over ground suitable for mechanized arms, it is obvious that all the principles of war can be more rapidly applied by a mechanized antagonist. The object can more easily be maintained, because mobility and security enable concentration, surprise and offensive action to be developed. Further, economy of force and cooperation are more readily gained on account of the enemy's inferiority in protected offensive power. Against such an enemy the task is really too easy to warrant much discussion. We should be like Charles VIII of France when he invaded Italy in 1494. Machiavelli tells us that he conquered that country with a piece of chalk. What did he mean? He meant that because Charles had so strong a train of artillery nothing could resist him—neither castle, fortified city, nor field army—all he had to do was to chalk on the map where he wanted to go and there he went. A mechanized army operating against a non-mechanized one will do exactly the same thing, it will be like pitting modern battleships against early nineteenth century three-deckers.

[Example, the overthrow of France in 1940.]

When both sides are partially mechanized, the governing principle is economy of force; consequently application depends upon a correct distribution of troops to ground, the non-mechanized arms seeking security from mechanized attack by operating over broken and enclosed areas, whereas the mechanized ones develop their mobility over the less difficult spaces, working at a distance but still in cooperation with the former, which should be looked upon as the tactical base of their movement. The mechanized arms should attempt to surprise the enemy by concentrating offensive power against his flanks or rear.

When both sides are fully mechanized, surprise, mobility and concentration against the objective are the governing principles, from which should be developed security and offensive power in cooperation, so that economy of force may result. As mobility will probably be the same for both sides, surprise becomes all important, therefore, command of the air is an essential factor.

LECTURE II

CHAPTER II

Fighting Troops, Their Characteristics and Armament

9. General

The difficulty in the last war was that we were at the end of an epoch, and the difficulty today is that we are at the beginning of another. During the First World War the defense grew so powerful that the infantry attack became impracticable unless led forward by tanks or strongly supported by guns. This war showed clearly the enormous holding power of the older arms, and it was this holding power which led to trench warfare. As I have already pointed out, the main characteristics of the newer arms are their mobility and protective power, and from these we can deduce that, so long as the present period of transition lasts, nothing but confusion will result from mixing the old and new. To combine tanks and infantry is tantamount to yoking a tractor to a draught horse. To ask them to operate together under fire is equally absurd.

[When I wrote this I had in mind infantry as commonly understood—namely, as footsoldiers whose raison d'etre is to fight their like. Motorized infantry equipped with anti-tank weapons are not infantry, they are anti-tank foot.]

On what principles should we separate these two, in order to make the most of each? The answer is to be discovered in mobility and protection. When the ground is such that tanks cannot move over it, or only with the greatest difficulty, obviously it is better suited for infantry, and in any case they will not be faced by tanks. When the protection is such that infantry cannot be attacked by tanks, as in a fortress or an anti-tank area, or are unlikely to be attacked by tanks, as when distant from the enemy's tanks or close to friendly tank forces, obviously their utility will continue. But in all places where they can readily be attacked by tanks, not only are they next to useless but a perpetual source of anxiety.

In areas where tanks can move freely we must divorce our minds completely from present-day tactics, for fighting will be very different. First, the bullet will be eliminated, anyhow as one of the principal weapons; secondly, the close cooperation which we see between the shell and the bullet will no longer be possible, for offensive power will be based on shells and armour-piercing projectiles and not on bullets. In a way, battle between tanks will take on a naval character, but with one great difference: ships fight on the sea and are based on the land, tanks fight and are based on the land only. What does this mean? It means that ports, harbours and coastal fortresses, though they cannot follow ships at sea, can, in the form of their land counterparts, follow tanks on to their actual battlefields. If at sea it were possible to move a defended harbour behind a fleet, naval battles would resemble closely what future battles on land will be like.

I have already in my last lecture touched upon this subject. I have mentioned the organization of Alexander the Great's army; his heavy cavalry were based on his phalanx—a slow moving fortress of pikes. Even today we see the same thing: the infantry are based on artillery; but in the tank battle there is no infantry, it is a tussle between self-propelled armoured guns, and what are these weapons to be based on? They should be based on what I will call the anti-tank troops—the modern representatives of the medieval wagon laager and the classical phalanx.

[Of anti-tank troops there are two main categories: (1) Artillery-anti-tank foot with anti-tank guns, or anti-tank gunners with self-propelled anti-tank guns; and (2) Engineers—mine-layers and anti-tank trap and obstacle constructors. The 25-pounder field howitzer-gun is an excellent anti-tank weapon.]

In my opinion, the future will reveal to us that a mechanized army, or formation, must be organized in two wings—a tank force for offensive power, and an anti-tank force for protective power. The first consisting of mobile machines, and the second of transportable anti-tank weapons which can follow up the first and establish a modern "wagon laager" in rear of it or in its vicinity. I should like to retain this picture in mind; for, as I proceed, I shall frequently have occasion to elaborate it. Rifle and gun to-day are replaced by tank and anti-tank, because armour has eliminated the bullet.

[This requires clarifying. The tank itself is the offensive anti-tank weapon, the self-propelled gun—shielded or unshielded is the mobile protective anti-tank weapon, and the transported guns, mines, etc., are the static protective anti-tank weapons. The dive bomber when used to attack tanks, is a combination of the offensive and mobile protective anti-tank weapons translated from the earth to the air.]

10. Infantry

In battles between armoured machines infantry can play no part worth their risk; but, as I have already pointed out, tanks are out of place in mountainous or thickly wooded country, and further—they are never likely to be so numerous as to be able to occupy an area and control it. As mountain and forest warfare cannot be overlooked, highly trained infantry, or rather light infantry, will be required for such operations, and as occupation of the enemy's country is in nine cases out of ten a necessity, and always a visible sign of victory, I have suggested for this work soldiers of the "second degree." We thus obtain two categories of infantry—first line and second line.

[Today three categories would be more correct—namely:

(a) Light Infantry for fighting in areas unsuited to tanks.

(b) Infantry of the normal type, but specially trained and equipped to occupy conquered areas.

(c) Motorized Infantry trained for guerilla fighting, or what I have called "motorized guerillas," but considerably more highly trained and organized than I considered necessary in 1932.]

The first armed for bullet warfare—that is, with machine guns and rifles—and the second, as I have already suggested, equipped to carry out the duties of pioneers and gendarmes. Police work, the protection of communications against raiding forces and the construction of fortified centres and defended depots in the conquered area are the main duties of the second line.

11. Cavalry and Mounted Rifles

Cavalrymen today are no more than mounted rifles, for the sword and the lance are not weapons, but instead the encumbrances of tradition. So long as the masses of the infantry march, the existence of cavalry is probable, but once these masses become motorized, even if only for logistical purposes, they will disappear. The reason for this is obvious; for though it is easy to motorize a man, it is unprofitable rather than difficult to motorize a horse. A cavalry division followed by hundreds and hundreds of horse transporters, superimposed upon its already large number of forage carts and trucks, would be a nightmare rather than a fighting formation.

Cavalry will die a lingering though natural death, but the cavalry idea will certainly not die. It will live on in the form of the motorized trooper, and the mechanized one as well.

As regards the former, a distinction must be made between the mortorized guerilla and the motorized cavalryman. The first is an irregular pure and simple, mobilized on the outbreak of war, and of special utility when working in his own country.

[As I have pointed out, he should be trained and organized to operate efficiently in all countries.]

The second, a highly trained scout is part of the first line organization.

[When the motorized guerilla is placed on a higher tactical footing than originally I considered necessary, the distinction between him and the motorized cavalryman lapses. The word "cavalryman" is now an archaism, such as are "man-at-arms", "grenadier," etc. Except, perhaps, for police work, cavalry is an arm of the past.]

When mounted in a light motor car, though these scouts may not be able to go everywhere a horse can go; they can cover far greater distances at far higher speeds—that is to say, they will accomplish in a given time far more than present-day cavalry can. This will fully compensate for minor deficiencies.

As regards the latter, mechanized cavalry, we have the armoured car which some soldiers think will eventually be replaced by a light tank, because this vehicle possesses a higher cross-country capacity. This may be the case; nevertheless, it should not be overlooked that mechanized cavalry will be in constant demand for distant reconnaissance work, for the attack on isolated posts, bridges, railway stations, aerodromes and headquarters; therefore the wear and tear of the machines will be considerable, and until the tank possesses a higher endurance than it does today, it would seem unprofitable to make this exchange. Further, it should not be overlooked that most of the targets mechanized cavalry will be ordered to reconnoitre, or attack, lie on roads, and that extreme speed will generally be demanded in these operations.

Another duty which this arm is likely to carry out is to form strong nuclei for the motorized guerilla forces. These, if supported by squadrons of mechanized cavalry, will be endowed with a far higher confidence in themselves; further, as they may be lacking in discipline, pillaging and other irregularities, which are so frequently amongst hastily raised bands of irregulars, can be better checked.

12. Artillery

As in the armoured battle the rifle bullet can play but a small part, radical changes may be expected in artillery tactics and, therefore, in artillery equipment. Today we have four main categories of guns—siege (including heavy and medium), mountain anti-aircraft and field. Of these the first three will be required, and we may accept it that they will be very similar to those of today. The fourth obviously will be different, because the present-day eighteen-pounder gun and 4.5 in. howitzer are designed to cooperate with infantry and not with tanks.

[Field guns intended to support tanks should be self-propelled.]

The tank, be it remembered, is nothing more than a field gun mounting. It is equipped with a gun of small calibre, such as the three-pounder, because this weapon is sufficiently powerful to smash through a comparative thin sheet of armour. As armour increases in thickness, probably the calibre of the gun will increase in size.

[This is what has happened, the 6-pdr. has replaced the 2-pdr.]

Do we require a more powerful gun to cooperate with the three-pounder? The answer to this question depends on tactics. In present-day fighting the eighteen-pounder and 4.5 in. howitzer cover the infantry attack from a distance and depend upon indirect laying. In the tank battle, except when smoke shells are used, such support is not practical; for it is out of the question to attempt to render assistance by firing into a melee of tanks, some of which may be moving at between 20 and 40 miles an hour.

[An example of this is the fighting south of Tobruk in November 1941.]

Smoke, as cover from view, is however important, and though vast clouds of smoke can be generated from a tank, the actual front of an advancing machine is best protected by smoke shells. Therefore we may conclude that a howitzer tank will be required.

[Such a machine, known as "a close support tank" was designed shortly before the outbreak of the war. Today I should prefer a 25-pdr. to a howitzer.]

Again, a tank may come up against a strong point—earth work, house, village, etc.—against which a three-pounder is a poor weapon; here, then, is another duty for the howitzer tank—namely, to bombard these points and cover the tank attack on them.

So far artillery support and cooperate in the tank versus tank battle. But this is only half the problem, the second half is the gun protection of the anti-tank base of operations—the mobile fortress behind the tanks.

As the bulk of its guns will be either hauled or transported, it is a great advantage if they are kept light, and as a small shell can knock out a tank, a small gun is obviously the right weapon.

[This no longer holds good, as tank armour has been considerably strengthened, from 15-mm in 1932 to as much as 90-mm in 1942.]

Therefore we may decide on the three-pounder carried in a cross-country vehicle with its armoured shield or cupola. This weapon will fire armour-piercing shell, and will depend for effect upon accuracy of fire.

Besides accuracy we shall want volume of fire, especially for close quarter fighting. And though it may be possible to invent a three-pounder automatic gun, on account of the difficulties of ammunition supply it would appear more practical to rely on an anti-tank machine gun firing a one inch or three-quarter inch armour-piercing projectile, for such a bullet will pierce an inch of armour at from 500 to 700 yards range.

[Such a weapon is now obsolescent if not obsolete.]

We therefore arrive at the following categories of guns in a tank formation:

(1) Tank wing: Three-pounders and field howitzers.

(2) Anti-tank wing: Three-pounders and anti-tank machine guns.

The projectiles being three-pounder armour-piercing shells, howitzer smoke shells, high explosive shells and armour-piercing machine gun bullets.

13. Armoured Units

As regards armoured machines, the first point I should like to note is the clumsiness of the official nomenclature of medium, light and close support tanks; because a name, whenever possible, should denote the function of the weapon.

[This nomenclature has now virtually lapsed. We have today two types of tanks—Infantry Tanks and Cruiser Tanks. The first are allotted to Infantry Divisions and the second to Armoured Divisions.]

It is as sensible to talk of expensive and cheap as of medium and light. It is the function which matters, and a function first originates as an idea, and around this idea should the machine be built as well as named. What then are the tactical ideas which should control invention? They are—finding, protecting and hitting; therefore we want machines which can find, which can protect and which can hit; in other words—reconnaissance tanks, artillery tanks and combat tanks.

As regards the first of these three categories, what type of machine should a reconnaissance tank be? The answer depends upon whether reconnaissance is to be distant or close. If the former, then we require a long machine, because speed in this class is essential, and a long machine will negotiate accidents of the ground better than a short one. As, generally speaking, it would be extravagant to use such machines for purposes of close reconnaissance, and as in this form of reconnaissance a less conspicuous machine is less vulnerable, a scout machine will also be required, a machine somewhat similar to our present-day light tank.

[Our original light tank was designed more from a fighting than a scouting point of view, and is now obsolete. I am still of opinion that a correctly designed distant reconnaissance tank is required.]

As regards the second category, it is likely to be divided into two classes of machines, an out-fighting and an in-fighting one. The former, as I have already mentioned, will consist of self-propelled armoured artillery mounting howitzers of sufficiently high calibre to give smoke shell as well as high explosive. The second class should, so I think, consist in what I will call a scout destroyer, a machine somewhat more powerful than a scout tank and carrying an anti-tank machine gun or semi-automatic small calibre gun.

[On account of increased armour, this machine is now obsolete. Instead, each combat tank should be equipped with an in-fighting and an out-fighting weapon, for example, a 6-pdr. and a 25-pdr. or guns of equal power.]

Such a machine, using volume fire, would be particularly useful in close-quarter fighting.

The third category—the actual fighting tanks—should, I think, consist of two classes of machines, a combat tank something like our present medium tank, and a pursuit machine which quite possible will coincide with the reconnaissance machine. The importance of the combat tank over the lighter machines lies in its heavier armour, greater power to cross obstacles, and above all in its higher moral effect. The terror which tanks aroused during the World War was well recognized by the Germans, but it was only partially realized by ourselves, and is today almost forgotten. The moral effect of a present-day light tank equipped with one machine gun is considerable, but of a sixteen ton medium tank moving at 20 miles the hour and firing one three-pounder gun and four machine guns is overwhelming. This is generally overlooked by advocates of light machines.

Of special machines it is difficult to see where they will end, for as armies become more and more mechanized a number of types will undoubtedly be required. The following seven machines will almost certainly be built: A water-crossing tank, an assault machine, a supply tank, a bridging tank, a gas tank, a mine-layer and a mine-sweeper. As regards water-crossing tanks, I am of opinion that all reconnaissance and pursuit machines should be able to propel themselves under their own power across water—river or sea. The assault machine I visualize is one which may be compared to our present experimental heavy tank, its main characteristics being not weapon-power, or speed, but armour, which should withstand all projectiles up to non-armour-piercing field gun shells.

[Today this machine is represented by the Infantry Tank, but in my opinion it should be still more powerfully armoured—possibly up to 100-mm.]

Before I turn to the next Section I should like to summarize the points I have now dealt with, so that my listeners may have a clear picture in their heads of the general organization of a tank formation or army.

It will consist of two wings, an anti-tank and a tank wing. On the line of march the first will follow and be protected by the second. At the halt the second will at once deploy and establish a protected area defended by guns, anti-tank machine guns, mine fields, and various types of works, in order to provide a haven of refuge not only for the services and such auxiliary troops accompanying the army, but also for the tank wing.

From it the tank wing will issue forth to give battle to its like, and to it will it retire should it be hard pressed or defeated. This wing will comprise reconnaissance tanks, scout tanks, artillery tanks, destroyer tanks and combat tanks, which will operate on somewhat similar lines to a fleet at sea.

14. Engineers

Turning now to the Royal Engineers, here again we may expect radical changes. At present the Royal Engineers are in the main field engineers; their duties including fortress building, bridges, road construction, water supply and demolitions, all of which are just as important to a mechanized army as to a present-day one. The Royal Engineers, I feel, should be brought into far closer touch with the mechanized arms than they are to-day. Not only should they occupy themselves closely with questions of assisting tank mobility and restricting it, but they should be responsible for all field repairs other than maintenance, which is the duty of the tank crews. The Royal Engineers should, therefore, be divided into two bodies of men—field engineers and mechanical engineers.

[A Corps of Electrical & Mechanical Engineers was created in 1942.]

The latter being organized in brigade and divisional companies, as the case may be, each equipped with a field store and the necessary mobile workshops for field repairs, the heavy repairs being carried out by the Ordnance Corps.

15. Aircraft

Though aircraft are totally motorized they are not mechanized. To armour an aeroplane adequately may not be impossible, but it is unprofitable as the essential features of this machine are speed and radius of action.

[This is incorrect, as most machines today are partially armoured.]

This arm, possessing a three dimensional movement, stands in a category by itself, faced by a future of all but unlimited possibilities; for when wireless control becomes more perfect, it is certain that we shall see the manless flying machine—the true aerial torpedo. When such a weapon is invented the whole form of war, on land and at sea, may once again have to be recast.

Today one limiting factor in air offensives is the difficulty in hitting small targets. What the aerial bomber looks for is something he cannot well miss—a city, an extensive depot, a long column of man or a fleet of transports. The larger the target the more profitable becomes the attack.

It may be said that to hit a column of infantry is by no means an easy thing to do. This is often true if hitting is looked upon as striking the men in the column. But the men have nerves, and these are "hit" every time aircraft are signalled; for a mere threat to attack will at once cause a marching column to disperse and assume a protective and, therefore, **static** order. This striking at mobility is a far more effective means of hitting than attempting to destroy a few lives, because, if persistently carried out, it must result in an ever-increasing demoralization of personnel and disorganization of plan.

In comparison to an infantry column, a mechanized column is far less vulnerable. Not only will it generally offer a smaller and more dispersed target, but a threat to hit will not necessarily, or normally, be followed by a halt. To damage it, is must be actually struck, and should one machine in it be hit, it is unlikely that any other machine will suffer. From this we see that one answer to air power is armour; armour on land has a more detrimental influence on air tactics than fire. A bullet, or a shell, may bring a single aeroplane down, but armour "strikes" at air power itself by denying moral effect to inaccurate bombing.

In aircraft organization the main tactical limitation is not to be sought in the air, but on the ground. There lies its "rear," which can be struck at from the air or from the ground, and of these two attacks the second is likely to prove so deadly that aerodromes and aircrafts depots, etc., will be kept as far back as possible, wereas those which are forward will have to be strongly fortified against tank and motor vehicle attack, so strongly that they are unlikely to be frequently moved. The less movable they are, the more vulnerable they will become to air attack.

From these few remarks we see that there is a close relationship between mechanization on the ground and motorization in the air; in fact the tank and the airplane are complementary machines, and for a long time to come the one is unlikely to be able to operate safely without the other. The airplane, by finding the enemy tanks, will enable the tank to attack or contain them and so protect the aerodrome. Without the tank the rear of the air is largely uncovered, and without the aeroplane the front of the tank is largely blind. Therefore we can deduce this fact: That in future warfare cooperation between tanks and aeroplanes is likely to prove far more important than cooperation between tanks and infantry. So important that we may see tanks and aeroplanes forming one force and infantry a completely separate force.

[The advent of the dive bomber has accentuated the importanco of this close cooperation.]

I will now turn to the purely protective point of view of the anti-aircraft problem.

There is little doubt in my mind that anti-aircraft appliances and artillery will grow so effective as to make it highly dangerous for aircraft to attack a strongly protected area. But how many areas can be strongly protected? The answer is "very few," and in the tug-of-war which will take place between industrial and political centres and military ones, it is the latter which are the more likely to receive secondary consideration. It would, therefore, seem that every effort should be made to restrict the number and limit the size of military objectives which offer good targets to the enemy's air force. In its turn this can only, economically, be done by cutting down the size of armies means substituting quality for quantity.

When it comes to the question of protection on the move, not only do present-day armies with their immense slow-moving trains offer favourable targets to air attack; but worried as they will be by guerilla forces, and constantly threatened by mechanized ones, efficient protection against air attack becomes all but an insolvable problem. The main reason for this is that the organization, control and protection of anti-aircraft guns on the move are far more difficult questions than at the halt, or in fixed positions.

Thus we see that, however reactionary military thought may

be, the influence of air-power on the older arms is likely to compel their replacement by such as are more self-protective, less ponderous and more speedily in movement; arms which can cooperate with aircraft instead of relying upon aircraft to escort them from position to position. Arms which need to be escorted are arms of a low fighting category.

16. Smoke

Smoke is an excellent shield against aimed fire, but it is a poor protection against volume fire. In an infantry attack it is obvious that, if the enemy's artillery O. P.'s are blinded, the attacker will be protected from artillery fire; but to put a smoke cloud down in front of the attacking infantry is almost useless, as all the enemy machine gunners need do is to fire into the cloud. Replace infantry by tanks and protection is effective, because failing an abnormally large supply of anti-tank machine guns, it is by aimed artillery fire that tanks are destroyed. In brief, smoke and armour are complementary protective agents, whereas smoke and woolen jackets are not.

As regards the creation of smoke clouds, there can be no question that, in certain circumstances, their generation from fast moving machines is far more effective and far more economical than by smoke shells. Literally for a few shillings a dense cloud a mile long can be emitted by a single tank, which if produced by shell fire would cost thousand of pounds. This does not mean that smoke-shells should be abandoned; for, as I have shown, they are likely to prove invaluable in the tank battle.

17. Gas

The influence of lethal and vesicant gases, as well as lachrymatory chemicals and toxic smokes, on the older arms is so fully appreciated that it needs no accentuation. A man in a gas mask is only half a soldier, if that. In a vesicant-proof suit of clothing he is worse off still, for he is three-quarters a diver and, consequently, no soldier at all. But as a tank is a metal box on tracks it can be made gas-proof, just as it can be made water-tight; therefore tanks which are gas-proof have little to fear from a gas attack, and can themselves carry large quantities of gas to attack with. For such attacks the method which suggests itself is to hitch on behind a tank a series of large drums containing liquid gas, and to drop them off by means of a quick release operated from within the machine, which will at once bring into play a time-fuze fixed in the drum. A present-day medium tank could easily haul ten two hundredweight drums, and, consequently, could release one ton of gas in a few minutes. As tanks can maneuver for wind, which shells cannot do, this form of attack is likely to become a deadly one, especially in outflanking operations.

LECTURE III

CHAPTER III

Strategical Preliminaries to Joining Battle

18. General Considerations

When in my first lecture I discussed "The Nature of War," I pointed out the importance of industry on military organization and of military organization upon strategy and tactics. As the machine gains control over civil life, so will it eventually gain control over military thought, which in recent history has never been less than a generation behind the civil thought of its day. Thus, though the steamship was introduced in 1812, it was not adopted by the Navy until about 1840; and though the percussion cap was invented in 1814, British infantry were not armed with a percussion musket until 1842. Had it not been for the first World War, there can be little doubt that we should only now be beginning to think in terms of the tank, and that we should be doing so is highly probable, for its origins are not to be sought in that war but in the civilization which gave birth to it.

The two great influences of machinery upon economic civilization are increased production and decreased labor. Whilst a little over a hundred years ago wheat harvested with a sickle and threshed with a flail required about forty hours of work per acre, to-day a tractor and combine will do this work in less than one hour. In manufacture it is the same, and this acceleration of work, due to the substitution of machine power for muscle power, will undoubtedly be carried into the military sphere, with the inevitable result that, as weapon-power increases, fighting man-power will decline, until mechanized forces will more and more revert to the numerical level of armies in the Middle Ages. Then 500 knights in armour were a large and powerful army, so will 500 tanks be in the future.

[This requires modification, because tanks are now reckoned by the thousand. Nevertheless, 500 is still a powerful force and is not often likely to be exceeded at any given time in any one area on account of the difficulties of control.]

A marked influence of this decrease in the size of armies, more especially as it concerns the combatant arms, will be the increase of flexibility of action and, therefore, of plan, which will be further accentuated by power to move across country.

[As combatant arms have decreased in numbers, their services in rear have increased. In 1914-15 the generally accepted percentage was 40 non-combatants to 60 combatants; today it is probably 50 to 50. Further, it probably takes five civil workers to maintain one military fighter or non-fighter in the field. In other words, an army of, say 4,000,000 men requires 20,000,000 civil workers (men and women) to maintain it, and of its total, half are combatants and half non-combatants.]

In the past, strategy has been woven upon communications, but what are communications today? They are not only roads and railways, but open spaces. Thus, in a flat land, such as most of South Russia, except for the rivers and forests there are no hindrances to the movement of tracked vehicles. Therefore in such areas it would seem probable that future warfare will approximate closely to warfare at sea.

[Still more true of Libya.]

As communications melt away into broad strategical tracks of country, possibly scores of miles in width, it is obvious that their attack will become more difficult. Again, because armies will be smaller and far more mobile, it will be easier to turn the flanks of a hostile force and attack it in rear.

[It should always be remembered that, when in these lectures I note the comparative smallness of armies, I am considering their mechanized forces and not the infantry mass or army of occupation.]

Both contentions are, I think, correct; yet it must be remembered that when an army attacks another in rear it does so to cut it off from its base of supply, and when it attacks its communications its object is identical. It does not really attack a road, or a railway, or an area, for its aim is to sever an army from its base. Should the future prove that this severance is so difficult as to become over risky, or unprofitable, the alternative will be to attack the bases instead of the communications. Bases are likely, therefore, to grow in importance, not merely the main base, but particularly the forward ones, and be it remembered the more bases a mechanized army possesses the higher will be its maneuverability; just as in warfare at sea the maneuverability and radius of action of steamships increases in direct proportion to the number of coaling stations, oil bunkers and defended harbours.

What does this mean? It means the increased value of field fortifications, not only to defend these bases of supply, but also to block narrow approaches, as once did medieval castles and strongholds. One thing is certain, the more mobility increases the more will fortifications increase.

[If mobility breeds stability, it does not follow that the fortifications of a mechanical age should re-duplicate those of a muscular.]

Field warfare always begets siege warfare; it did so in the last war, and it will do so in a war between armoured forces, but with this difference: whereas in the First World War fronts were fortified, in a war between armoured forces areas will be so instead. By this I do not mean that an area will be walled in by protective works, but that these works will be dotted about in such a manner as to place the communications of any force entering the area in danger. The difference between these two systems is that of the Great Wall of China compared to medieval castle defense.

[The mistake in the Maginot Line was to relate its defensive power to an offensive approaching obsolescence. Had it been broken up and scattered in depth, it would, I think, have fulfilled its purpose.]

The one stops an invader; the other does not, instead it establishes a series of pivots, really bases of supply, between which the invaded can maneuver in comparative safety.

Turning from these general considerations to the influence of mechanization on our Imperial strategy, a moment's thought will lead us to see how beneficial it is likely to be. First we are one of the grestest industrial nations in the world, and are, consequently, admirably equipped to produce all the war machines we may require. Secondly, apart from assisting in maintaining law and order in Great Britain, the two duties of our army are: (1) To assist some ally in a war on the continent of Europe, and (2) To defend the frontiers of the Empire. As regards the first, our army has never been organized for such a war, and though many of our training manuals talk of it being trained for a war of the first magnitude, it is certainly neither recruited nor organized for such an eventuality. In 1914 we were able to send six regular divisions overseas in a matter of a few weeks; today (1932) we are not prepared to send two divisions overseas under several months. Now, if instead of thinking in divisions, we were to think in terms of one small highly organized mechanized army of some 30,000 officers and men carried in armoured and unarmoured vehicles, and think of this force as **always mobilized,** then not only could we render immediate assistance, but assistance in a form most desirable to our ally. A force of this size, though numerically only a fifth of our Expeditionary Force of 1914, would be out of all proportion more powerful, especially should the enemy still rely on a conscript horde—that is, upon numbers of men instead of perfection of weapons.

[An hypothetical example is the following: If in May, 1940, Lord Gort had had under his command 45,000 fully mechanized troops—that is, three armoured divisions—would not he have been better placed than he was with his army of 300,000 non-mechanized troops? Had he had 90,000, I go so far as to suggest that the probabilities are the Germans would have been defeated. Six armoured divisions would have required some 2,000 tanks, which would have cost about £40,000,000, or three days war expenses.]

To turn to the second of these two problems. Shall we want an immense army? We may, but one thing is certain: No British Government will maintain such an army during peace time. Should we require such an army, then it will have to be raised after war is declared. Therefore, being compelled during peace time to maintain a small army, we should not leave a stone unturned to make it as powerful as possible.

What does power demand? Above all things rapidity of movement, the rapidity of a fire-engine which can extinguish a conflagration before it catches hold. The advantage of motorization and mechanization are that they reduce space by economizing time. In other words, the more rapidly we move the smaller becomes the bulk of the area we are called upon to defend. Strategically, time and space are relative, and as the history of was has shown again and again, a handful of men at a certain spot at a certain hour is frequently a far more powerful instrument of war than ten times the number on the same spot twenty-four hours later. If you can move five times as fast as your enemy, then, whilst the military hour will remain 60 minutes for him, it will be reduced to 12 minutes for you, and every mile will become less than two furlongs.

If from our own problem we turn to the one which faces most continental countries, we find that the requirements of mechanization clash with those of conscription. This I have already noted in my first lecture. The result is likely to be that change over in these countries will be slower than in those in which professional armies are raised. Nevertheless, in the end it is as certain; for not only are masses of infantry completely out of date, but against mechanized arms they are useless.

[Infantry as infantry and not as anti-tank foot.]

When they do mechanize, the influence on their grand strategy will be profound; for as mechanization will be costly, not only will their armies be reduced in size, but their governments will have less to spend on their air forces and navies, and, therefore, will have radically to modify their present strengths.

[This demands qualification, because the German finance system introduced in 1933 was based on labour and not on cost as normally understood. It was related to working parties and not loans. Man power and not money power was its chief concern.]

19. Strategical Reconnaissance

Mobility in finding and in hitting are complementary forces, which in another war will constantly influence each other's worth. For example, it will often prove useless to risk aircraft in finding an enemy unless immediate advantage can be taken of his discovery, and more particularly so in strategical operations. The aeroplane and the marching man are incompatibles, for their rates of movement are so different. Here is presented to us another reason for speeding up movement on the ground, and when this movement is speeded up, marked changes are likely to take place in the initial operations of a war.

[It would have been more correct to have written "revolutionary changes." Examples: the German attacks on Norway, Holland, Belgium and France.]

In pre-mechanized warfare the normal plan was to push forward large forces of independent cavalry as soon as war was declared, the main object of these forces was to gain liberty of movement for the infantry masses which followed them. As this independent cavalry and these masses had to be fed, it was useless to move either forward unless their lines of supply—formerly roads and more recently railways— were adequate for this purpose. Consequently, supply controlled strategy with an iron hand, and when a theatre of war possessed only one railway, as happened in the Russo-Japanese war of 1904-05, strategy was enormously simplified.

This simplification is now becoming a thing of the past, because though strategy is still based on supply, the grip of supply on strategy has become far less tenacious. Whereas the supply-radius of an infantryman is very restricted, by which I mean that he can move only a few miles before he has to be fed and also rested, that of an aeroplane runs into hundreds of miles and that of a tank into scores. This enhanced power to cover distance, coupled with the facts that mechanical vehicles can carry in bulk a reserve of supplies, and, if provided with caterpillar tracks, are largely independent of roads, has rendered strategy far more flexible, the grip of supply being replaced by that of information; for unless accurate knowledge of the enemy is gained, tactical mobility will suffer in proportion.

The seeking of information by aircraft will undoubtedly lead to battles in the air, but that these engagements will be directly sought at the opening of a war is, in my opinion, improbable, unless both sides are equally matched, or possess some technical advantage which they believe to be unknown to their enemy. What is more probable is that determined threats will be made in certain directions in order to cover strategic reconnaissances in others. Such operations will require extremely fast machines rather than large numbers. I am of opinion that the bulk of an air force will not, unless circumstances are extremely favourable, risk battle, and by battles I do not mean skirmishes and duels, until decisive operations on the ground take place.

[Generally speaking, this has been borne out. In May, 1940, the R. A. F. would seem to have avoided battle, not because it could not have attacked, but because of the land situation. It was only after the French army collapsed that the German air force set out to attack the British Isles. Again, this attack was abandoned directly war was carried into Russia.]

It is the overwhelming blow which above all others paralyzes an enemy's will, and in the future the object will undoubtedly be simultaneously to strike such a blow on the ground and in the air.

[Which means cooperation between air and ground until the land battle is won.]

If I am right in this, then, at the opening of a war, the main advantage of air power will not lie in decisive action but in reconnaissance, because superiority of information means the gaining of the initiative, and not until the initiative is gained can a definite plan of action be decided on.

[Also, as the war has shown, the pulverization of the enemy's air fields.]

Whilst air reconnaissances are taking place, the mechanized forces are more likely to be distributed over a wide area rather than to remain concentrated; for with motorized and mechanized arms it is far easier to concentrate small though scattered units than to move masses of machines from one place to another. To secure this wide distribution it will be necessary to push well forward of it along the whole front a screen of armed motor cars and anti-tank and aircraft detachments, the object of which will be in order of precedence: (1) To supplement the strategical reconnaissance of the air force and discover the enemy's approach; (2) to occupy positions of strategical and tactical importance, and (3) to delay the enemy's advance.

Behind this screen, which should work to plan with the forward aircraft, the armies will advance on a wide front until information becomes so definite that a concentration in a given direction can be ordered. When this is possible, success will very largely depend upon mobility.

[The German-Polish campaign of 1939 is an example of this.]

20. Special Considerations in the Case of an Overseas Campaign

In recent years overseas expeditions have become increasingly difficult, so much so that today they are almost impracticable against a well equipped enemy, because he can within a few hours of a hostile fleet being signalled concentrate by motor vehicles considerable forces against any threatened point on his shore line. Few of such points are likely to exist, as a modern army cannot disembark on any open stretch of beach. In fact it is extremely doubtful whether a present-day army with its immense impediments of supply columns, hospitals and work-shops—a veritable travelling city—can disembark anywhere outside a well furnished port. To effect a surprise landing is not only difficult but almost impossible, and without surprise a disembarkation becomes, in nine cases out of ten, a suicidal undertaking.

[This was in part, at least, corroborated by the Dieppe raid carried out in August 1942. The St. Nazaire raid, which was smaller, would seem to have come as a surprise. The Japanese invasions—except that of the Island of Singapore—belong to a different category—operations against undefeated beaches and poorly roaded countries.]

Mechanization offers us a definite, practical and comparatively easy solution to this problem. The first tank was invented to bridge the bullet swept zone between two systems of trenches, it was as it were a mobile trench which crept over static trenches. The landing problem is the same, it is to extend the protection afforded by the ship over a no-man's-land of water, and thence inland—that is, **into** the enemy's position. Its solution is the same, or at least very similar, it is a self-propelled amphibious tank, not launched into the sea from any kind of vessel, but from a tank carrying vessel from which it can crawl into the sea as easily as aircraft can take off from an aircraft carrier. An amphibious tank was suggested as far back as December 1917, and after the war an experimental floating tank was built; yet it was only last year (1931) that a true self-propelled amphibious machine was produced, and though its armour is too light, the problem of the water crossing tank may be considered as solved. It is no longer an idea but a fact, and I see no reason why in the future the bulk of the tanks of an army should not be able to cross rivers and considerable stretches of the sea under their own power.

With such a machine, launched from a tank carrier, surprise once again becomes possible, and the difficulty of bridging the gap between ship and shore is overcome. The landing can be effected on any normal beach, and at several places simultaneously. Air protection will be less necessary, the supply problem vastly reduced, and unless the enemy can meet the invading tanks by their like, it should generally be possible by a converging movement to sweep down on the rear of a port or harbour, effect its surrender and so gain a place of disembarkation for the main invading forces and their impedimenta to be landed at.

Besides definite landing operation, such as I have just considered, an amphibious tank will prove an invaluable weapon for coastal raids and for crossing rivers. To-day three-quarters of England is in easy range of continental aircraft, and as in another European war the bulk of the enemy's aircraft engaged in bombing operations against England will be based on coastal aerodromes, or anyhow aerodromes not far from the coast, and that here also will be located the submarine bases, it is obvious, if these aerodromes and bases can be attacked by raiding forces of tanks that they will demand anti-tank defenses, and so draw strength away from the main land forces.

Power to cross rivers without bridging them is so obviously

an advantage as to need no accentuation. Not only do they then cease to be obstacles of importance, but may actually be turned into lines of patrol; for an amphibious tank can "steam" up and down them, and not only frustrate hostile bridging operations, but attack its like in the water, as in the American Civil War the gunboats attacked each other on the Mississippi.

Finally, the utility of this machine is so great, and particularly to us an island power, it cannot be doubted that before long its needs will be fully appreciated, and that it will form part of our normal war equipment.

LECTURE IV

CHAPTER IV

Battle

21. General

The aim of strategy is to clinch a political argument by force instead of words. Normally this is accomplished by battle, the true object of which is not physical destruction, but mental submission on the part of the enemy. The idea that an enemy must be destroyed is only legitimate when it leads to a profitable state of peacefulness. Thus, should the enemy be an uncultured barbarian, his removal may, in certain circumstances, be a benefit to mankind; should he, however, belong to a cultured race—that is, should he be useful to the world, then his slaughter, even if unavoidable, must be considered unfortunate, and if avoidable —criminal.

It is important to realize this, because destruction as the end of battle is the common doctrine of to-day, and in the last war it led not only to the defeat of Germany and her allies, but to the defeat of the object of the war itself; for the destruction wrought during it so unhinged the nations mentally and morally that the peace established on this neurosis has proved itself no less destructive.

The reason for this fatal dogma was ignorance of the true object of war, which is to establish **a more perfect peace.** Its popularity was due to mental lethargy; for to think like a wild beast is easier than to think like a philosopher. Its unappreciated cause was the impotence of infantry in the attack, and its accentuating factor—the masses of infantry employed; for throughout military history the horde has always proved itself to be the most destructive instrument of war.

If the true object of war is to be gained, then this destructive mania must be eliminated. This means that war must be raised, little by little, from the cock-pit of the physical struggle into the spheres of intellectual and moral conflict. In other words, that generalship must largely replace brute force, and the moral attack—the attack on nerves—the physical dog-fight, or the attack on bodies.

[This theory was highly developed by Herr Hitler, see "Hitler Speaks," Herman Rauschning, pp. 15-21.]

On account of the low mobility of infantry masses, and the constant rest and rationing these masses demanded, method ousted originality and absorbed the art of war. In fact the instrument was such that it was incapable of withstanding the stresses and strains of original movements—flashes of generalship—and instead could only be manipulated according to a ritual in which generalship disappeared and was replaced by quarter-mastership.

To reinstate generalship the instrument must be changed, and to replace, not entirely of course, but in greater part the physical attack by the moral attack—that is, by the unexpected or surprise attack—the instrument must be highly mobile, also it must be as small as possible, for the smaller it is the less extensive will be its supply organization, and the less extensive this organization is, generally speaking, the more readily can it be extemporized and the better can it be protected; therefore the less vital and difficult will become its security.

With a small well balanced and flexible army, an army which can stand the stresses and strains of unexpected movements, and an army which is not permanently tied down to a fixed line of communications, generalship can be developed into a high art, and battles into works of art and not merely daubs of blood. It is here that motorization and mechanization come to our assistance, enabling us to re-establish generalship by changing the organization of armies. The more important of these changes are the following:

(1) **Armour.** As I have already shown, the bullet is at the base of all our present troubles. It prevents infantry closing, it forces artillery to keep well in rear of the infantry front, and it denies shock power to cavalry. Armour can defeat the bullet; therefore a tank can replace infantry in the attack, be-

cause it can ignore the fire power of infantry in the defence. Though tanks cannot be attacked by infantry, infantry can readily be attacked by tanks in all localities over which mechanized arms can move, unless infantry are amply provided with weapons and defences which tanks cannot face. This means that, when confronted by tanks, infantry are deprived of their mobility, they cease to be field troops, and therefore, should be used as fortress troops instead.

Though armour can defeat the bullet, it can be defeated by the shell. This does not mean that armour is rendered valueless, but that its value is relative to projectile power. As it can keep out the lesser projectiles but not the greater, this means that artillery will more and more preponderate over infantry, and as artillery will have to be protected against bullets, guns must be armoured. Therefore, the answer to the tank is the tank; therefore the present-day infantry battles will be replaced by mobile armoured artillery battles, and though in these battles the armour used may be no protection against the shells fired, it will have to be maintained in order to prevent the bullet coming back into its own.

[This has been proved correct in all open country.]

(2) **Mobility.** In field warfare protection loses the greater part of its value unless coupled with mobility, and every change in mobility, more particularly protected-mobility, carries with it a change in tactics. When to protected-mobility is added ability freely to abandon roads and move across country, these changes are radical, because existing linear tactics will be replaced by area tactics—this I have already noted. In area tactics the front of an army will no longer so completely protect its rear services and its line of communications as to-day. Attacks will take place in areas and not against lines, they may come from any direction. Therefore power to surprise is vastly increased, therefore the moral attack will grow in importance, its aim being to effect disorganization through demoralization rather than through destruction.

[The classical example is the collapse of France in 1940. The anonymous author of "The Diary of a Staff Officer" writes: "The French General Staff have been paralysed by the unorthodox war of movement . . . the 1914 brains of the French generals responsible for formulating the plans of the allied armies are incapable of functioning in this new and astonishing lay-out."]

As rear attacks will become more and more prevalent and ever-threatening, it logically follows that all-round protection will have to be established not only when at rest, but when on the move.

[The Russians would seem to have tumbled to this, with the result that many of the German rear attacks during 1942 proved abortive.]

Rear depots will have to be fortified and the mobile rear services included within the logistical formations, or the outpost line, and landing grounds as well as anti-aircraft defenses will have to be protected against raiding forces. More and more will an army become a static fortress when it halts and a mobile one when it moves.

(3) **Ground.** The connecting link between armour and mobility is, as I have already shown, ground. As area warfare grows in importance, more and more does it become necessary to consider the use of armoured machines in relation to ground. As economy in fighting will only be gained when each weapon is related to the nature of the ground, and as for a long time to come the two main fighting arms are likely to be tanks and infantry, a battlefield may be divided into two areas—namely, tank areas and infantry, or anti-tank areas, or those over which tanks can easily operate and those which afford ground protection to the infantry. The first will generally be open, flat, or undulating country, and the second, wooded, swampy or mountainous.

From the above analysis we are able to build up in general terms the nature of future battles: The gradual replacement of infantry by tanks—mobile armoured artillery.

[Though, so long ago as 1925, we had a well-designed self-propelled 18-pdr. gun, in the 1941 autumn campaign in Libya we were surprised by this weapon—a German 75 mm gun mounted on a Pz.Kw.3 chassis.]

The consequent subordination of the bullet to the shell; the increasing importance of surprise on account of protected cross-country mobility; the consequent importance of all-round protection, and lastly—the increasing importance of ground and tank obstacles. On these factors will future generalship be built.

22. The Forward Movement from the Area of Concentration

According to the plan decided upon, movement from the area of strategical distribution, which will replace the present area of concentration, will take place either immediately before the declaration of war, or simultaneously with it. If the frontiers of both belligerents are well protected by anti-tank defenses, no attempt to break through these is likely to take place—that

Fig. 1 Fig. 3 Fig. 2

Diagram 1. Column Formations.

is to say, if they are strongly garrisoned—until the main forces are ready to move. Should, however, they be inadequately defended, not a moment is to be lost in pushing forward large numbers of motorized troops equipped with anti-tank weapons, to occupy and fortify tank approaches, and such positions the holding of which is essential to the fulfillment of the plan. These troops should be rapidly reinforced by strong detachments of tanks and armoured cars; the first to protect them and the second to reconnoitre ahead of them in conjunction with aircraft.

On account of the ever-present danger of flank or rear attacks, column formations should be in echelon, as shown in Diagram 1. Which type is adopted will depend on the nature of the country and the number of roads; for it must be remembered that, though tanks and similar machines possess high cross-country powers, whenever possible they will move by road if only to avoid obstacles such as streams, woods, walls and fences. Should an impassable obstacle exist on the right flank of an advancing force, then the formation shown in fig. 1 may be adopted; if no such obstacle exists and roads are few, then that shown in fig. 2 and if roads are plentiful—as in fig. 3. In each case the transport and rear services will march either between the formation and the obstacle, or within the formation. Generally speaking, large masses of machines should be avoided, as speed of movement varies in inverse ratio to size. A number of small formations advancing on a wide front and in supporting distance of each other will accelerate movement without impeding concentration; for in nine cases out of ten power to move rather than to mass is the foundation of concentration.

23. The Approach March

Under cover of the advanced motorized troops the approach will take place, not so much in an order of march as in the order of a general distribution; that is in a number of small columns using many roads and at times moving across country, and so distributed that all-round protection on the move is maintained, instead of a few large columns protected by special guards and detachments. When the frontier is reached and the enemy's country is entered, though the width of the front of the advance may not change, in order to increase their power of attack and defense, the columns are likely to close inwards, forming strong groups on the main lines of approach, each group being protected by a swarm of motorized guerillas.

I consider that the normal formation of all the groups should be either that of the arrow-head, (fig. 3 of Diagram 1) or open square (fig. 2 of Diagram 1—the Napoleonic lozenge), the group nearest to the enemy acting as the advanced guard. If the open square formation is adopted, then (see Diagram 2), directly **a** comes into contact it should attempt to hold the enemy, whilst **b** and **c**, or parts of these forces, maneuvre against his flanks.

The important point to remember is that the order of distribution (the area order), is the order of battle—that is, troops must be **distributed** as they are likely to be required and not merely marched in this order.

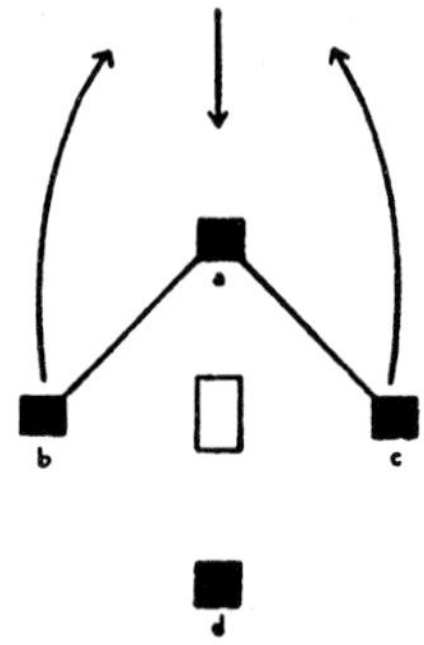

Diagram 2. Action on Contact.

24. The First Contact with the Enemy

The first contact with the enemy will obviously be between the contending air and advanced motorized forces, and for some time after hostilities have begun, I think, it will be the exception rather than the rule that either of these forces will gain a decision. In my opinion, their role is more to bewilder the enemy than to defeat him.

[This I now consider to be incorrect. Instead, whenever possible, the advanced guard action should aim at immediate and overwhelming decision. It may be compared to the first charge of armoured knights.]

The influence of their attacks is, therefore, likely to be moral rather than physical.

Under the protection of this artifically generated fog of war, the advanced forces (**a** group in Diagram 2) will move forward as rapidly as possible with the one idea of engaging the enemy, not to annihilate him but to fix him, not so much by hitting him as by maneuvering him into difficult ground or against an obstacle. Infantry are normally pinned down by fire, but fire will seldom pin down a mechanized force, it will have to be boxed-up by maneuvre.

[This is what the German armoured forces so frequently failed to accomplish during the 1941 campaign in Russia. For the reasons why, see my "Machine Warfare" pp. 171-173.]

A point worth noting here is that the formation of groups must be as flexible as possible. As the enemy will generally have to be fixed by maneuvre, group **a** may have to move in any direction at the shortest of notice, to which movement groups **b**, **c** and **d**, will have to conform. It is no longer a question of "follow my leader," but of maintaining a correct position in an area, one side (group **a**) of which is rapidly changing direction. If the earth were like a billiard table, conformation would be easy enough, as it is at sea, but normally it is very different, being broken up by woods, mountains, valleys, etc.

[In other words, tactics are largely governed by topography. This is why in Russia they are very different from what they are in Libya.]

25. The Plan of Battle

Distribution is plan in the bud, action is plan in full bloom. The rapidity with which action should unfold itself may be taken at between five and ten times that of present-day fighting; therefore each hour we have to-day to plan in, modify a plan in and issue orders and instructions in, will be reduced to from twelve to six minutes. A fixed idea is out of the question, the idea of the plan must be flexible—that is, it must embrace a number of alternative actions. There can be no formal method of attack or defense, though, as a general rule, it may be accepted that an attack which threatens the enemy from all directions and a defense which can frustrate such a series of attacks, are the ideals to aim at. The reason for this is that battles in the future are likely to become more and more area operations and not merely positional ones.

Outside enemy action, the main factor which will influence the plan is the nature of the ground—the facilities and difficulties it offers to mechanized movement. Other factors are wind and sun; wind with reference to gas and smoke, and sun with reference to visibility and fire.

[General Rommel—so it has been reported—prefers to attack late in the day, in order to have the sun behind him. Dust and dust storms must be added to this list.]

In the future plans will have to be based more on movement than on offensive action; by which I do not mean that the offensive will become less important, but that its effect will be the more dependent on correct and rapid movement. Once the opportunity to strike offers itself, movement must be immediate and by signal and not by operation order.

[It would appear that in Libya the Germans have appreciated this.]

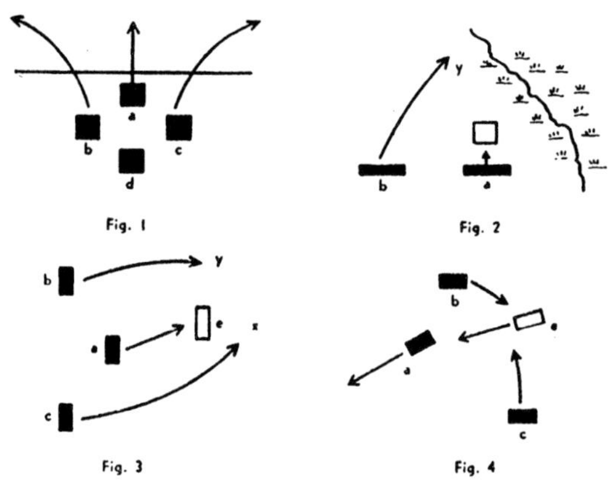

Diagram 3. Forms of Attack.

To lay down forms of attack is difficult, other than the general rule that the enemy should be boxed-up in an area before he is decisively attacked.

[At Dunkirk the British and French were boxed-up on three sides. The campaigns in Russia and in North Africa have shown how difficult this maneuvre is. For its tactics see my "Machine Warfare," pp. 103-104.]

The following formations are, however, worth considering:

(1) **Penetration.** If the enemy is holding a defensive position which cannot be turned, then his front will have to be broken. In such operations the open square formation is a suitable one, group **a** effecting the penetration, groups **b** and **c** enlarging the

gap, whilst **d** forges ahead and attacks the enemy's rear (see fig. 1, Diagram 3).

[Such a formation would appear to have been used by the Germans when in May, 1940, they broke through the French defenses at Sedan.]

(2) **Single Outflanking.** If the enemy is maneuvred diagonally towards an obstacle which he cannot cross, or can only cross with difficulty, a single envelopment may encircle him. Thus, in fig. 2, if **z** is a swamp, then whilst **a** presses him in front, if **b** can move to **y** his line of retreat will be cut off.

[Up to the summer of 1942, when, on account of the Qattara Depression, an open southern flank ceased to exist, this has been the outstanding maneuvre of all campaigns in Libya.]

In such an operation it will at once be seen that it is the maneuvre and not the attack which is the foundation of victory.

A similar operation can be carried out by a retirement, the enemy being lured against an obstacle, or induced to enter an area flanked by two and similarly attacked.

(3) **Double Outflanking.** When obstacles do not exist, a mechanized group must be used in their place. Thus, in fig. 3 of Diagram 3, whilst **a** moves against the enemy **e, c** maneuvres towards **x, a** driving on to **c** whilst **b** moves on **y.**

[This maneuvre is more fully dealt with in my "Machine Warfare," pp. 98-104.]

Again, a similar operation can be carried out by retiring, **a** falling back and drawing **e** towards it, whilst **b** and **c** remain out and clinch inwards as **e** advances (see fig. 4).

It will be seen that these operations are converging rather than outflanking ones, as distinct forces are employed and not one force. This being so, a point worth noting is the importance of hiding these forces so that they may surprise the enemy.

[An outstanding example of hiding a considerable force of mechanized and motorized troops in so open a country as the desert of Western Egypt is General Wavell's deployment on the night of December 7, 1940—that is, immediately before he launched his attack on Marshal Graziani.]

This means three things: (1) The necessity of gaining local

command of the air; (2) the part which woods, valleys and dips in the ground will play as hiding places, and (3) the use of motor guerillas to search these out.

26. Deployment for Battle

Deployments, such as take place to-day, of columns into lines will be replaced by expansion or contraction of distribution into fighting formations, this change over taking place as the enemy is approached. With marching columns of troops the main difficulty is that rapid movement cannot be made in extended order, the columns deploying to fight and ploying to march. This difficulty is largely overcome by mechanized troops, for though when advancing and retiring they will if possible use roads, road movements are not essential.

27. Organization in Rear

The present problem of supply in the field is tactically a simple one, because the front of an army normally protects its rear. In area warfare this protection is to a large extent non-existent, because flank and rear attacks are easier. The result of this is that a reversion is likely to take place to the supply system which held good during the seventeenth and eighteenth centuries—namely, defended field depots with convoys working in between.

[Though there has been little information on the working of the German supply system in Russia, it would appear that, on account of guerilla warfare, the field depot and convoy system has been resorted to in certain localities.]

In armies largely dependent for their mobility on gasoline, it is obvious that the protection of the convoys will become all-important, and that, consequently, special mobile line of communications protective units will have to form part of the field army. The system will probably be as follows: A base depot strongly protected against air attack, with fortified depots linking it to the field army. Each group, or corps, will draw its supplies from an advanced depot, and will be responsible for the protection of its columns. Like the old independent cavalry, the corps supply columns are likely to be duplicated, so that the full one may move with the corps whilst the empty one goes back to refill. The depots on the line of communications will require protective troops of their own.

Though, normally, supplies will be sent forward to the depots by road or rail, operations may at times demand cross-country supply; therefore the organization of the line of communications will probably include one or more cross-country trains—high power tractors which can haul over ordinary field land upwards of a hundred tons.

[It is my opinion that, had the vehicles of the German supply columns, attached to their armoured and motorized formations, been tracked instead of wheeled, the Russians would have been decisively defeated before winter set in in 1941.]

These trains may be compared to the horse transport units now abandoned, but which were maintained as a reserve for several years after trucks were in use, in case truck supply broke down, or became inoperative.

Another point worth noting is the likelihood of the reintroduction of foraging. This was common a century or so ago, but in modern times was given up, because the size of armies became so great as to render the gathering in of supplies a negligible factor. Not only will motorized and mechanized armies be smaller, but gasoline and oil will become so necessary to the very life of such forces that every endeavor will be made to seize all stocks found in the enemy's country.

[From several accounts it would appear that, in May and June, 1940, large forces of German tanks and vehicles lived almost entirely on undestroyed French fuel.]

Practically every town and village in Western Europe now contains one or more gasoline pumps; and it cannot be doubted that these will be used in war in order to supplement supply and so economize transport.

28. Positions of Commanders

The correct position for every commander is at the point of greatest importance, a point which is constantly changing its position.

ARMORED WARFARE

[General Rommel's several successes would appear to have owed much to this. The office-general—unless his office is a tank—has no place in the armoured battle; he belongs to static warfare.]

Thus, when preparations are in progress, it is where they can best be supervised; when an advance is made—as near as possible to the head of the advance; once the battle is launched—at the most suitable point of ground and air communication, and when a disaster takes place—right forward with the shattered troops, or immediately in rear of them.

As mechanized formations will be comparatively small, in an attack (generally speaking) their commanders should lead their troops into action. In fact their safest place will be with their men and not behind them. An admiral in a battleship is safer than in a dinghy behind his fleet; so also must the commander of a mechanized force be in a tank and not sitting on a shooting-stick miles in rear of his command.

LECTURE V

CHAPTER V

Information

29. Information—General Principles

Information is the foundation of battle, and battle is the main tactical action which takes place between armies; consequently information varies according to the nature of the forces engaged, their armaments, means of movement and protection; in fact their composition and general organization. For example, whereas a horde army is tied down to road and rail communications, a mechanized army is to a far greater extent free from them, for it can move across country and be supplied by cross-country vehicles. Further, it is less conspicuous, or rather its strategy is less conspicuous; for, on account of its speed of movement and its greater independence from communications, it can move on a broader front and in greater depth. It covers an area and not a series of lines, and can move from cover to cover at greater speed, or more rapidly change its position under cover of darkness.

Until aircraft was invented, one of the great difficulties in horde warfare was to discover the enemy. Once introduced, this became almost a matter of routine, for size and slowness of movement prevented the horde being lost once it was found. In the case of a mechanized army, though there should be no great difficulty in finding tank and other forces in spite of the smallness of their size and their higher speed, there will be a real difficulty in maintaining contact with them. It will be a game of hide and seek, unless command of the air is definitely won, and even then the ability to shift position during night time will often out-wit air reconnaissance.

[An example of this is General Wavell's first Libyan Campaign when by a night movement he completely outwitted Marshal Graziani.]

Another difficulty, more especially at the outset of a campaign, is that motorized and mechanized troops will be much dispersed, and that, therefore, their general direction will not be easy to gauge; and yet another, that whereas in horde warfare strategy, tactics and war doctrine generally are formal, governed as they are by communications, in mechanized warfare they will be less so, certainly until our present system of mapping has been radically modified.

Strategically, existing maps are of importance because they show roads and railways, tactically because they show rivers, mountains, etc.; but a mechanized army wants more information than this, it wants to know what areas suit its movements and which do not. The slope of hills, the surface of the ground, the banks of rivers, the nature of woods, all affect its movement. Generally speaking, it wants to know where tanks can move easily, where with difficulty and where not at all.

Yet another problem which will face a mechanized army is reporting the enemy's movements in time. A greater one—to be able to take advantage of this information; for as mobility increases so will the duration of its importance diminish. It is often imagined that such inventions as aircraft and wireless communication will lift the fog of war. It is true they may lift certain corners of it; but, in my opinion, the bulk of the fog will remain just as dense, for increased mobility will cause situations to change rapidly, and constant and often conflicting information from air and ground will bewilder a commander as fully as complete information renders him cautious. Finally, I think, that negative information will grow in importance, because to know that the enemy is not in a certain area will often be more important than positive information of a conflicting nature.

30. Reconnaissance—General Principles

Whereas a few years ago the means of reconnaissance were limited to cavalry and infantry patrols, to these must now be added (1) airplanes, (2) stationary balloons, (3) motor vehicles and motorcycles, (4) armoured cars, and (5) tanks. As the means have been vastly extended so must the allocation of duties be more precisely worked out, in order that the highest economy may result. Though formerly there were only two spheres of reconnaissance, an outer strategical and an inner tactical one; the first the area of cavalry formations and the second of cavalry and infantry units, to-day there are three: a deep strategical sphere, a shallow protective sphere and in between a wide tactical sphere. The first is the area of aircraft, the second of tanks, cavalry (so long as they hold the field) and infantry—on foot, in motor vehicles and on motorcycles—and the central one of motor and armoured cars. The importance of this last sphere can scarcely be exaggerated; for if I am right in supposing that the motor-guerilla will play an important part in another war, this central sphere will be occupied by a veritable army, a fast moving swarm which will not only search the area of advance, picket bridges and tactical points, block roads, etc., but will fight off the enemy swarm and so clear the area of advance.

[This happened in May and June, 1940, the German motor cyclists and reconnaissance units representing the motorized guerillas: but the French had none to meet them.]

Should the enemy reinforce his motor-guerillas with tanks, then his opponent will have to do likewise; but, generally speaking, I consider the tank will be too valuable a weapon for guerilla warfare.

Within the guerilla swarm will have to move a number of aircraft communication stations, or offices, to which air information can be sent back, and from which armoured car patrols will, as occasion demands, be sent forward to reconnoitre for the swarm and to beat back the enemy motor-guerillas.

Within this swarm the army will move and rest, the swarm thinning out in front and thickening on the flanks as the enemy is neared, until, when contact is gained, except for a few patrols, the front will be cleared.

Though, as I have already noted, the means of reconnaissance have increased, it must not be supposed that the actual work has become much easier. The smallness of the main forces and the ubiquity of the motor-guerilla forces will render finding difficult and judgment perplexing.

[Again, this was the case with the German motor cyclists in France. They moved so quickly that no sooner were they reported at one spot than they had occupied and were reported at another.]

Rapidity of movement will enable changes of direction to be more readily made—a wooded country, a foggy day, or a night move may upset all calculations.

31. Reconnaissance from the Air Before Battle

In mechanized warfare the first principle of air reconnaissance is that it should be unceasing. From the moment war is declared air information of the enemy must be gained, and if contact is not to be lost it must be maintained night and day. The call on aircraft will, therefore, be exceedingly severe, and, until command of the air has been gained, this will render persistent attacks on the enemy's cities and centers of industry unlikely, unless at the outset of the war his opponent is vastly stronger than he is.

[With reference to this: the Germans did not launch their air attack on Great Britain until they had first settled with France, and they ceased their air attacks on British cities shortly before they attacked Russia.]

In the maintenance of contact, motor-guerillas and armored cars will play a leading part. What the aeroplane discovers these weapons will hold. Thus, should an enemy force be signalled from the air as being at **a**, then the object of the motor-guerilla and the armoured car will be to swarm round this locality and occupy all approaches and exits, in fact to box the enemy up so that in whatever direction he moves his movement will be discovered—or covered.

[This was typically the action of the German motor cyclists in France during May and June 1940.]

Should the enemy be strong in aircraft, in order to gain liberty of air action, it may become necessary to employ a stronger reconnaissance formation of all mechanized arms which can move forward independently of the main army, seize and occupy a land ground, fortify it with anti-aircraft artillery and searchlight and so establish a protected base for air action. Such a formation would in fact be a moving fortress. It would consist of a strong force of anti-aircraft artillery protected by tanks and mobile anti-tank defenses, covered by an outer ring of motor-guerillas and armoured cars.

The system which prevails today is that of aircraft moving from the rear of an army over its front to attack their targets. Though this system will undoubtedly remain the normal one, in the attack on the enemy's cities it may be necessary to push the air base out to the front, or more probably to flank, of the army—that is, separated from it—in which case it must be self-contained and self-protected.

32. Reconnaissance from the Ground before Battle

From what I have said about aircraft, it will be realized that reconnaissance on the ground will have to be as unceasing as in the air, and that these two forms of reconnaissance are interdependent.

With the motor-guerilla swarm I have already dealt, but so far I have said little as regards armoured cars and tanks, the duties of which I will now briefly enter into.

If the swarm is delayed by enemy motor car attack, this inevitably well mean that the main army in its centre will also be delayed. As actions between motor-guerillas will generally be frontal, to relieve the inward pressure caused by the enemy guerillas, the surest means will be to attack them in rear or flank by reconnaissance tanks or armoured cars—more probably the latter. The method I suggest is explained in Diagram 4: **a** is the main army surrounded by **b** the motor-guerilla swarm, **c** and **d** are two forces of armoured cars. As **a** and **b** move forward, **d** and **c** do the same, but outwardly in two large circles, **d** clearing the area into which **a** and **b** are about to move, and **c** circling round the outer edge of **b** to attack in rear, or flank, any enemy cars which may attempt to drive in or penetrate **b.** When **d** has arrived at **e,** it will, directly **a** and **b** are well in area **f,** circle out for the next bound, whereas **c,** now at **d's** original position, will once again carry out its circling duties. In actual fact the circles will not, of course, be regular as shown in the diagram, but will follow the road system immediately outside **b.**

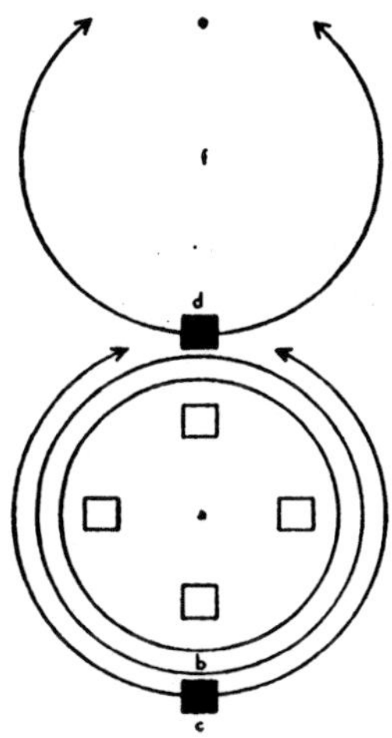

Diagram 4. Methods of Ground Reconnaissance.

Another important reconnaissance is that of the ground itself. In order to distribute a mechanized force correctly, a commander must have detailed knowledge of the types of ground over which his machines are going to move; therefore ground reconnaissance should not simply end with a general description of the ground, but with one which compares the ground with the various arms which may be called upon to move over it. This work, I consider, should be carried out by the reconnaissance tanks, these machines moving forward into area **f,** as **d**—the armoured cars—move round it. Not only will the armoured cars be able to supply information to the reconnaissance tanks, but these tanks will form a central reserve to them, should they be attacked by tanks.

As reconnaissance must be continuous, it should be kept up at night time, by which I do not mean that full movement should be maintained. This is out of the question, as men and machines must be rested: but that reconnaissance forces must not lose touch with any enemy forces discovered during daylight. This, as I have already explained, is best done by picketing them.

33. Reconnaissance during Battle

Once battle is engaged and the main army comes into action, the motor-guerilla swarm should concentrate on its flanks and operate against the enemy's flanks and rear. As this work is likely to prove highly dangerous on account of the proximity of the enemy, the enemy's mechanized forces, the main reconnaissances should be carried out by reconnaissance and scout tanks, the motor cars and armoured cars moving farther and farther out on the flanks to watch all approaches and to signal back the advance of any hostile machines.

Though, during battle, it is of the utmost importance that the maximum of information regarding the enemy should be gathered in, this is likely to prove of little value unless possible enemy's actions have been foreseen and prepared for. This does not mean, should the courses open to the enemy be six in number, that six complete schemes have to be prepared, instead one scheme with six equally good answers.

[It would appear that the fundamental defect in the British plan of attack in Libya in November, 1941, was its lack of flexibility. It presumed that were General Rommel surprised—as he was—he would retire. As it happened he did nothing of the sort, and the result was that General Cunningham's plan broke down and a series of confused actions followed. As the Cairo Special Correspondent of "The Times" reported on November 25: "There are no definite lines and no fixed plan, but merely masses of death-dealing tanks maneuvering, attacking and counter-attacking—in short, chaos."]

Unless this is done, the chances are that either an army will be frittered away in detachments, or that so strong a reserve will be held in hand that hitting power will be detrimentally affected. To take full advantage of any of these possible courses of enemy action, as well as to eliminate such as the enemy does not intend to adopt, reconnaissances must be scheduled accordingly, so that intimate touch may be maintained between possible enemy maneuvres and the actual distribution. If the enemy's reserve forces cannot be observed from the ground, they must be kept under constant air observation, the reconnoitring aircraft maintaining contact not only with command headquarters, but also with a special force of motor and armoured cars which will operate according to their instructions.

Besides these various reconnaissances, it must never be overlooked that it is the duty of every unit to reconnoitre as well as to fight and protect itself. This is doubly the case in mechanized warfare, for frequently units will be far spaced and movements will be rapid. To-day reconnaissances are comparatively slow and elaborate. A force proceeds on a mission, then reports in full to command headquarters, at which the report is digested and new or amended orders issued. Now it must be remembered that with machines moving at possibly 20 miles the hour, every half hour may carry an operation into a totally different area, and that, therefore, for fighting units to await orders is out of the question, yet they must operate to plan. How can this control be established?

I think a solution is as follows: All maps should be squared and each square numbered, and if all are tank maps, that is they show where tanks can go, can go with difficulty and cannot go, so much the better.

[The aim of this system is not to maintain secrecy, but instead to accelerate rapidity of movement and action.]

The plan is made out with its alternatives and every fighting commander knows these. The operation is launched and the machines disappear into the blue. Now what are the points a commander wishes to know so that he can control the operations? The position of each of his units, the position and strength of the enemy, the action of each of his units, the action of the enemy, and the condition of each of his units. There are other points, but these will suffice for example. Now put these into code: the map squares are numbered and each unit has a code call—Ex, Zw, etc. Actions: A = attacking, B = holding, C = retiring,

D = moving, etc. Strength given in words. Condition: E = strong, F = indifferent, G = bad. Now as to a message from front to rear or centre: "Ex 100 A En 101 fifty D 98 F" or "1st Battalion in Square 100 attacking enemy in square 101, his strength estimated at 50 machines moving towards square 98, his condition indifferent."

At command headquarters, in the open, or in a tank, the commander will also have a squared map set out like a chessboard, each square having several small holes in it. By his side will be a box of coloured, numbered and lettered plugs, or pins, such as are used in the game of cribbage. The board will be set at the opening of the operation, and as messages come raining in by wireless, a clerk will move the plugs from square to square, changing their numberings or letterings as information demands. As the operation proceeds, the commander will modify his plan by a somewhat similar code, sending out his code orders to units, or groups of units, and leaving it to the initiative of the men on the spot to carry them out. Thus, he may wireless out: "Ex Ey A Ez Fx B Fy Fz D 213 Gx D 209," which would mean: "1st and 2nd Battalions attack enemy, 3rd and 4th Battalions hold him, 5th and 6th Battalions move to square 213 and 7th Battalion to square 209."

In this system the two main points of importance are: (1) The commander must be provided with constant information; (2) the subordinate commanders must act on their own initiative with reference to an idea and not an explanatory order.

34. Information from Prisoners, etc. and from Captured Documents

As information is only of military value when it is up to date, and as in mechanized wartime may often be divided by five or six, it follows that much of the information called from prisoners and captured documents will lose its value. Again, as code and verbal orders will prevail over written and explanatory ones, it is likely that private soldiers will know less and less about the operations they are engaged in, or what the ideas underlying them are. More and more will the actual men become parts of the machine, whereas their officers become parts of the commander's brain.

[This would appear to be the basis of General Rommel's system of command.]

35. Secrecy Regarding Information

As the value of information is influenced by the speed of tactical movement, so also is that of secrecy. Before the opening of an operation the necessity for profound secrecy is obvious, which, as I have shown, can in part be maintained by the breadth and depth of the initial distribution. In horde warfare, because movement is comparatively so slow, secrecy is maintained throughout a battle; but in actions between mechanized units, once close contact and fighting take place, except for the movements of reserves, outflanking forces, headquarters and supplies, a wise, bold and confident commander will cast secrecy to the winds and rely on open code signals in place of complicated cyphers.

[In the 1941 campaigns in Libya, whereas the British command adhered to issuing orders in cypher, the Italo-German command did not. The result was loss of time and confusion.]

There is another advantage in this—namely, the bewildering and misleading of the enemy by code messages which are not to be acted upon. Remember this, that in future warfare ruse and stratagem are going to play all-important parts, for the quicker fighting moves, the sharper must be the wits of the general. It is the fox in man, and quite as much so as the lion, which will win through.

LECTURE VI

CHAPTER VI

Protection

36. General Protection

The reintroduction of armour and the general adoption of the motor vehicle are daily changing the whole outlook of tactics. The first provides direct protection against the bullet, the second renders this protection mobile; hence a new protective base is given to offensive action, a new and stronger foundation which will support a far more powerful tactical edifice. In bullet warfare the ground was the protective element, and to it was added extensions between riflemen in order to reduce the target. In mechanized warfare mobile armour replaces static earth and speed of movement from one position to another adds an increased value to extensions. Whereas, heretofore, the attacker has had to halt in order to fight, now, like a ship at sea, he can fight when moving.

The value of the motor vehicle is two-fold: it can be used with or without armour, and so, as I have shown, it gives rise to two classes of troops—mechanized and motorized—the first are the protective sword of the second, and the second the protective shield of the first.

As mechanization presents unlimited opportunities for surprise, and as in area warfare there is no well defined front to be threatened, general and local protection will become increasingly important. To-day an army is nowhere safe; it can be attacked from great distances by aircraft, and from considerable distances by motor and armored cars. When within twenty or thirty miles of its enemy, it may also be attacked by tanks, therefore local protection is daily becoming more vital, so much so that in future wars the posting of a few pickets and sentries will be quite insufficient.

Camps a hundred miles and more in rear of the front, ammunition dumps, railheads, etc., will have to be protected not only against aircraft but against raids carried out by armoured cars and motor cars.

[In the various Libyan campaigns there are to be found many examples.]

The importance of protecting headquarters, transport and artillery lines against air and ground attack is daily becoming more noticeable, and more and more is it necessary to establish these in anti-tank areas, and to hide them away from observation.

In attack, the first protective problem is the correct appreciation of ground to weapons. The second is the proper distribution of artillery, by which I mean that the guns should be placed so as to be in a position to assist whichever arm is to play the more important part. Thirdly, anti-tank weapons should be distributed in such areas in which the enemy is to be held back; and fourthly, tanks should be concentrated in those areas in which it is intended to develop the offensive. The whole attack should be organized on a protective plan and founded on a protective base, the offensive being launched from that base.

Protection in defense is very similar to protection in attack; but, as the initiative is with the enemy, it is of the greatest importance that as strong a reserve as possible is kept in hand. The defensive screen, which is to hold up and delay the enemy, should mainly consist of anti-tank weapons, behind which a reserve of tanks is held to counter-attack the enemy in flank, or rear, at short notice.

[The use made by General Rommel of his anti-tank weapons during his May-June, 1942, Libyan campaign was masterly. His tanks were followed into battle by his anti-tank guns, so that should his tanks meet with strong opposition, they could retire under cover of the guns, and then, should the guns be in turn threatened, the tanks could support them by advancing against the flanks of the enemy attacking tanks. Cooperation was, therefore, intimate. In October, 1942, when General Montgomery attacked, it was the reverse, because Rommel's anti-tank weapons were static instead of mobile, also the maneuverability of their supporting tanks was restricted by the minefields in which the anti-tank strong points were embedded. In this battle Rommel's mistake was to mix static and mobile tactics. Had he kept his anti-tank weapons mobile, they could have maneuvered behind his minefields, when close cooperation could have been established between them and his counter-attacking tanks. As it was, both were destroyed in detail.]

If the protective screen can hold up the hostile tanks, it should be remembered that the enemy will almost certainly turn towards his opponent's flanks and rear, hence in these directions also must preparations be made to meet attack.

PROTECTION WHEN ON THE MOVE

37. Mobile Forces

As in mechanized warfare the flanks and rear of a force are the points which will be constantly threatened, to march a body of troops in one long column is to ask for trouble, not only because its flanks increase with its length, but because it offers a good target to air attack. Obviously small columns, such as brigade groups, are more suitable, and, as I have pointed out in a former lecture, they should march in echelon with their strategic flank thrown back, or else in arrow-head formation, or open square. Should one brigade group be attacked in force, it should assume the defensive until one or more of the other groups can be sent to its assistance.

When we turn to protective guards, the introduction of the motor-guerilla swarm will considerably modify their duties. Instead of acting as independent bodies they will form strong nuclei within the swarm, protecting it when required, and at times breaking through defended areas and other obstructions in order to keep the circle of cars mobile and intact.

ADVANCED GUARDS

38. Strength and Composition of an Advanced Guard

The advanced guard of a mechanized force will undoubtedly consist of all classes of machines—namely, reconnaissance tanks, scout tanks, destroyer tanks, combat tanks, artillery tanks, bridging machines and armoured cars. To it will most certainly be attached aircraft for reconnaissance, and probably motorized infantry for occupying and holding operations. Well ahead of it will move aircraft and armoured cars, and on its flanks motor-guerillas, linking it up to the flank guards and to the rearguard and so completing the protective circle.

Unlike the present advanced guard it will not move from tactical feature to tactical feature. Instead the swarm with its guards, of which the advanced guard is the leading one, will move from tactical area to tactical area, in which tactical features are but isolated points, islands, so to say, within a confined sea, or lake.

39. Advanced Guard Commander

Because the advanced guard is no more than a strong link within the protective circle, though it must possess a commander of its own, it would seem necessary that the whole circle should be under the command of one man, the commanders of the various guards being subordinate to him; for, unless this is done, disorganization through independent action may result. I consider, therefore, that there must be one circle commander under whom will come the four guard commanders as well as the commander of the guerilla swarm. His position will vary according to circumstances, but, generally speaking, he will accompany the guard which is immediately threatened, sometimes the advanced guard and sometimes one of the others.

40. Action of an Advanced Guard

Once the plan for an advance is issued, the first problem of the circle commander is to work out a series of movements **by areas,** and to issue provisional instructions for the occupation of each in turn, in which the work of each group of arms is laid down. These instructions should include a general distribution of forces, aircraft and armoured car reconnaissances the action of the swarm and the action of the guards.

As regards the advanced guard, its main problem will be to keep the protective circle moving in the desired direction. A distribution which suggests itself is: armoured cars well out to the front, followed by a vanguard of reconnaissance and scout tanks and a main guard of combat and reconnaissance tanks, in rear of which moves the anti-tank wing. Should the enemy be met with, he will immediately be pushed back, and if this is not possible, then held by the vanguard, the tanks of the main body moving outwards and the anti-tank wing brought up to consolidate the vanguard area and so establish a strong base of operations. From this base, one, or two, outflanking attacks can then be made, supported on their flanks by motor-guerillas.

If, instead, the guerilla forces on the flanks of the advanced guard are held up, the advanced guard should slow down its forward movement and detach a force, or forces, to operate against the flank or rear of the enemy. In either case the idea is the same—namely, rapid outflanking and the avoidance of a frontal attack, and still more so of an assault.

41. An Advanced Guard in Retreat

In the retreat of a mechanized army pursued by a similar force, the main danger lies in being headed off. Therefore, when the situation points to such a possibility, the advanced guard, using its anti-tank wing, supported by tanks, should form what I will call a funnel formation.

[This, it would appear, is what General Rommel did during the fighting round the Cauldron in the Libyan Campaign of May, June, 1942. Breaching his enemy's minefield, he pushed his tanks forward, and as he did so he brought up his anti-tank guns to form two protected flanks in rear of them.]

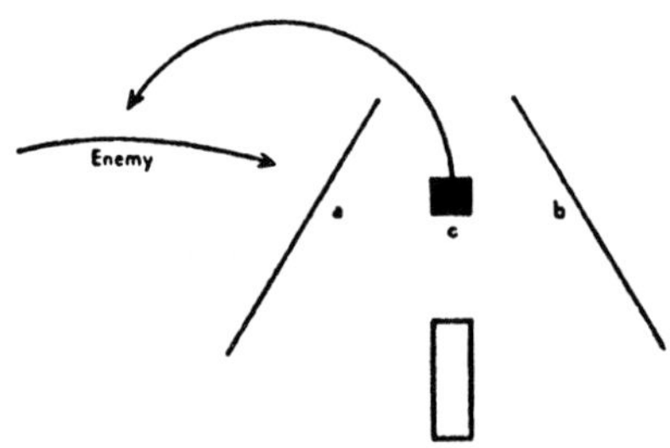

Diagram 5. An Advanced Guard in Retreat.

That is to say, it should throw out on its flanks two protective forces, (**a** and **b**), holding within the area between them a mobile force (**c**) ready at the shortest of notice to move through the apex of the funnel and take in flank or rear any enemy forces operating against its sides (see Diagram 5).

42. Protection When Deployed

In the case of a mechanized force, deployment, as generally understood, is a misnomer; for, as I have already explained, a mechanized force does not so much deploy from one formation into another, as expand, or contract, its original formation. This contraction, or expansion, will take place within the protective circle, the main object of which is to relieve the main body from protective duties whilst on the move and at rest. The further the enemy is away the larger is the area of the circle likely to be, but the nearer he is approached the more will it have to contract, the guards moving slightly ahead of the outer circumference of the swarm the better to protect it.

FLANK GUARDS

43. Composition and Action of a Flank Guard

In area warfare the composition of a flank guard obviously should resemble that of an advanced guard, because there is no fixed front, and at any moment such a guard may be called upon to carry out the duties of an advanced guard and, like

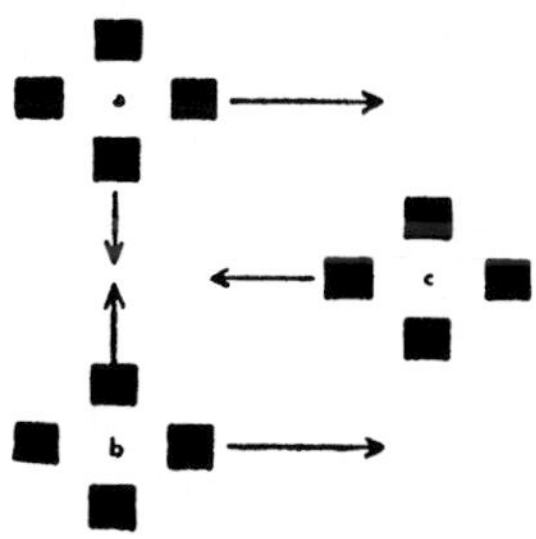

Diagram 6. Protection Against Flank Attack.

an advanced guard, it is nothing more than a strong nucleus within the protective circle. Therefore, its action will be similar to that of the advanced guard—namely, to beat off a determined enemy attack, or by attacking the enemy in flank prevent the swarm being driven in.

Should several distinct mechanized formations take the field, then, should they move in echelon, they will automatically reduce the likelihood of flank attacks. For example, in Diagram 6, **a** protects the front of **b, b** the rear of **a, c** the right flanks of **a** and **b,** whereas **c's** left flank is protected by **a** and **b,** and **c's** front and rear very considerably so by these groups which can take in flank an enemy attacking **c** in front or rear.

When a force halts, its flanks if possible should rest on natural obstacles—anti-tank areas.

REAR GUARDS

44. Actions of a Rear Guard to a Retreating Force

A rear guard to a retreating force should operate like an advanced guard to an advancing one—that is, it should move from area to area; but it must not be looked upon as a detached

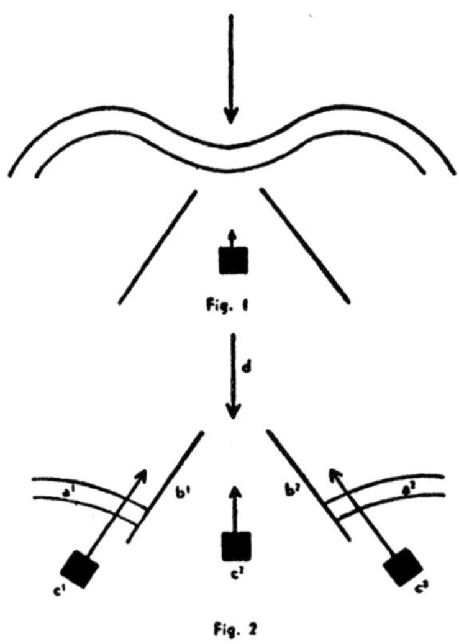

Diagram 7. Action of a Rear Guard in Retreat.

force, for whether in movement or at the halt, in advance or retreat, it remains part of the protective circle which must never be broken. The only exception to this rule would appear to be when an army passing through a defile is compelled to move in a linear formation. In this case the protective circle ceasing to exist, a detached rear guard becomes necessary.

The formation which a rear guard should assume is again that of a funnel, the guerilla swarm working outside it and maintaining contact with the enemy as long as possible in order to delay his advance.

[It was because General Rommel formed, as we have seen, an anti-tank funnel in rear of his tanks east of the Gazala-Bir Hakeim minefield (May-June, 1942) that, when his tanks had to fall back, they retired with little loss. The funnel protected their withdrawal like two shields—one on each flank. In fact the funnel was little else than a special form of laager.]

Diagram 7 illustrates a typical rear guard operation, fig. 1 showing the first phase and fig. 2 the second. In fig. 1, the swarm is still out but is being driven in; in fig. 2, the swarm is shown by the letters $\mathbf{a^1}$ and $\mathbf{a^2}$, $\mathbf{b^1}$ and $\mathbf{b^2}$ are the anti-tank wing of the rear guard, and $\mathbf{c^1}$, $\mathbf{c^2}$ and $\mathbf{c^3}$ the mobile forces. Should the enemy—**d,** penetrate between $\mathbf{b^1}$ and $\mathbf{b^2}$, $\mathbf{c^2}$ can advance and engage him whilst $\mathbf{c^1}$ and $\mathbf{c^3}$ attack him in rear and flank. Should the enemy attack $\mathbf{b^1}$, then $\mathbf{c^1}$ can operate against his right flank whilst $\mathbf{c^2}$, or part of that force, operates against his left flank. The secret of this particular formation is: (1) It establishes a defensive base from which the mobile forces can operate, (2) it induces an enemy to offer a flank to counter-attack.

45. Action of a Rear Guard in Close Contact with the Enemy

In mechanized warfare one of the supreme dangers is that a sudden reverse may result in a rout; for it is easier to lose control over fast moving machines than over slow moving troops.

[Notable cases of this are the routes of the British in the Sixth Libyan Campaign (May-June, 1942) and of the Italo-Germans in the Seventh (October-November, 1942.]

In medieval times a cavalry charge was frequently followed by the overthrow of one side or the other; nevertheless complete routs were infreqent, because each army of knights generally established immediately in rear of it a wagon laager held by infantry.

[The knights were wiser than the moderns. Should the present war have taught us anything, surely it is the vital tactical necessity of a properly constituted anti-tank wing or mechanized laager.]

These laagers were in fact mobile fortresses into which the knights retired when defeated. I do not suggest that in its entirety so simple a method of avoiding a pursuit can be introduced, but I do think that a mechanized attack should, whenever possible, be launched from an anti-tank base—that is, an area of ground easily defensible against tanks, or one occupied by the anti-tank wing. If this is done it should not be difficult to break off an engagement, and retire behind the anti-tank defences and refit.

The maintenance of a reserve is another important point, for, without an adequate one, a disengagement will become next to impossible. When used for this purpose, the reserve should not be drawn up in rear of the troops to be withdrawn, which is normally the case with infantry, but on one of their flanks, so that as the enemy advances he will be threatened in flank.

46. Demolition Schemes and Expedients for Delaying an Enemy's Advance

Unless demolitions are arranged for beforehand, it may be accepted that as a rule little or no time will be available to prepare them.

[In France in May-June 1940, it would appear that the most sketchy preparations were made; whereas in June-November, 1941, the Russians are reported to have systematically devastated the country they retired over.]

The blowing up of bridges will undoubtedly delay an enemy; but, should he be equipped with amphibian tanks, the delay will not be for long. A better system of delay would appear to be the construction of anti-tank posts and defenses well in rear of the retreating force. The value of these I will discuss in another lecture.

[Such was the Weygand plan of June, 1940; but it came too late, for the French armies were already in a state of dissolution.]

47. Rear Guard to a Force Advancing

With a force advancing through a friendly country a rear guard is likely to play a passive part. It will be within the guerilla swarm and will be protected by it; but directly the enemy's country is entered the likelihood of a rear attack must always be kept in mind, therefore the rear guard must be ready for immediate action.

When an army advances, the rear guard is likely to be large, for as a protected area of communications will have to be established, this guard will be accompanied by such troops as are required to garrison the field depots and provide escorts for the convoys which will work in between them. This problem I will deal with in my next lecture.

LECTURE VII

CHAPTER VI (Continued)

Protection

PROTECTION WHEN AT REST

48. General Nature of Outposts

Obviously protection when at rest will be influenced by the increased mobility of armies. At present, outposts are mainly formed by the infantry, and are distributed in order to protect resting forces from cavalry and infantry attack. They are placed sufficiently far out to gain time for the resting forces to get under arms. In motorized and mechanized warfare this picture will change, for outposts must be prepared to meet incessant armored attacks, which at any moment may be followed up by a determined thrust by armored troops. Because both these forms of attack are highly mobile, the outpost line will have to be pushed further afield and be prepared to beat off tank attacks.

As in such warfare the object of an outpost line is to keep a selected **area** clear of the enemy, the word "line" is really a misnomer, and should be replaced by that of "circle" or "circumference." Obviously the right troops to carry out this work are the guerilla swarm supported by the four guards—the advanced guard, rear guard and two flank guards. The general distribution is given in Diagram 8, **a** representing the resting army and **b** the guerilla swarm; the guards are lettered **c, d, e** and **f.** From the guards, or the swarm, should move out armored car patrols to a depth of some 20 miles—**g,** and beyond these, constant aircraft observation should be maintained, **h.** Thus, we obtain three protective circles—airplanes, armored cars and motor guerillas; the depth of the entire area depending on the proximity of the enemy. If near, the area included will be small, if distant—then large.

Immediately the air patrols report the position of the enemy, armored cars, or failing these motor vehicles, will be sent out to gain contact with him, and to picket all approaches between his position and the outpost area. As he advances they will fall back, and, as he approaches circle **b,** they will not retire within this circle but remain out on his flanks. Meanwhile, as I explained in my last lecture, the guards will make ready to take in flank any enemy force which may attempt to penetrate circle **b.**

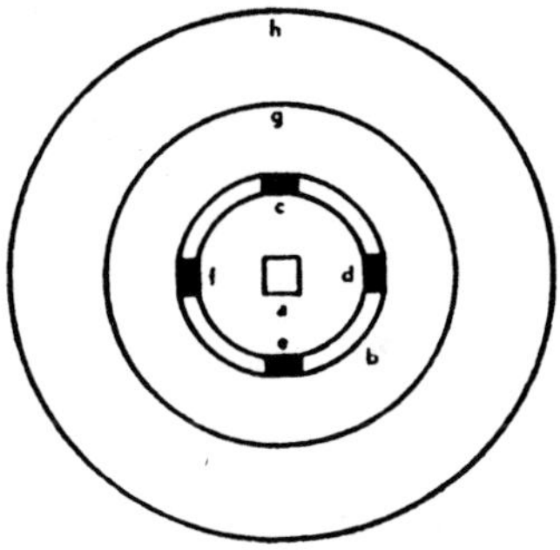

Diagram 8. General Distribution of Outposts.

From this it will be seen that protection at rest is in principle almost identical to protection on the move, the only difference is that, when at rest, an army can better protect itself against aircraft attack, because its anti-aircraft defenses—guns and lights—will be stationary.

49. Factors Affecting Protective Measures

The main factors which will influence protective measures are: (1) The mobility of the enemy; (2) the nature of the country; (3) command of the air, and (4) ability to picket the enemy effectively. If superiority in the air is assured, then the armored car patrols will be able to operate freely, and this freedom will increase in direct proportion as the enemy's power of movement is lower than his adversary's.

As regards the nature of the country, an ideal area to rest in is one encircled by rivers and hills, for then the enemy's motor guerillas will be restricted to a few avenues of approach. It is these forces, especially when operating in an enemy's country, which will generally give the most trouble.

[Whether Russian guerillas have at any time been motorized, I do not know. Nevertheless, they and their counterparts in Yugoslavia have given the Germans much trouble.]

The attack of an outpost area by mechanized arms during hours of darkness is likely to be the exception rather than the rule, because the control of such operations will always be difficult; but it must not be overlooked that enemy's tanks may approach during night time and lay up in woods, etc., until dawn. Should such hiding places exist in the neighborhood of the outpost area, they should be occupied by detached posts.

50. Protective Measures when the Enemy's Main Forces Are Not within Striking Distance

Should the enemy not be within striking distance—that is, should he be far away, well over a hundred miles—then, as I have already pointed out, the protective circle, whether at rest or on the move, should expand. In this case reconnaissances should be pushed ahead until the area occupied by him is discovered, when all roads leading out of it should be picketed. When far away, the only likelihood of attack will be by aircraft; consequently the deeper and broader the distribution of the resting, or moving, force, the easier will it be to conceal it.

Unless command of the air is absolute, which is seldom likely to be the case, it should be accepted as a rule that an enemy **is always within striking distance,** for today bombers exist which have a circuit of action of 1,200 miles.

[Of course much greater now.]

The fact that armies will never really be outside striking distance by aircraft and seldom so by motor vehicle has vastly increased the importance of the protective problem. It may almost be said that whereas the central idea of horde warfare is the cultivation of offensive power, that of mechanized warfare is the opposite. Protection will force itself more and more to the fore, not only in the form of armor, but in that of fortifications and defensive distributions.

[This has been clearly illustrated in Russia. The so-called "Hedgehogs"—fortified areas—have and are playing an increasingly important part.]

51. Protective Measures When the Enemy's Main Forces Are within Striking Distance

This problem I have already in part dealt with. It differs from the last one mainly in the size of the area occupied by the army, and also in whether this area is in the enemy's country or not. This point I have also mentioned, but here I will examine it more fully as it introduces another phase of the protective problem.

It is obvious, if I am right in supposing that motorized guerillas will play as important a part in future warfare, as I believe they will, that their most profitable sphere of action is in their own country, for not only will they be among friends but their supply problems will be far easier to solve.

When the enemy's country is penetrated, local conditions will rapidly begin to tell against both the security and mobility of the invading army, which, unless it is distinctly superior to its adversary's, will seek to gain a decisive battle in the neighborhood of the frontier. We may, therefore, expect that in the future frontiers will be systematically protected against motorized and mechanized attack, and that, consequently, prolonged operations of an indecisive nature will take place on them until a definite advantage is gained.

[Not by a wall of works, like the Maginot Line, but by a network of great depth.]

It also means that, when this advantage is gained and one side penetrates its enemy's territories beyond his anti-tank zone, not only will the invading side have to move within a closely guarded circle of troops, but also entrench when at the halt, as Julius Caesar did two thousand years ago, and for similar reasons. What will be the nature of these defenses?

They will not be trenches as we knew them in the last war, which were dug to protect the defender, for this type of protection is now being replaced by armor, which not only protects the defender but the attacker as well. They will be an extension of the same idea which gave birth to wire entanglements.

These obstacles were erected to impede the attacker's mobility: similarly in anti-tank defenses it is mobility which must be attacked—the advancing machines must be stopped.

Where will they advance? This is far from being an unanswerable question, for a large scale map prepared for mechanized warfare will show at a glance the approaches and areas the enemy's machines are likely to use.

To halt them and destroy them the following defenses suggest themselves: Mine fields which will replace wire entanglements and a network of strong points which will largely replace entrenchments.

[It has been reported—with what truth I cannot say—that the fall of Tobruk in June, 1942, was mainly due to a lack of mine protection.]

These strong points will probably take two forms: (1) Small redoubts or concrete pill-boxes, and (2) small portable bullet-proof, and possible shell-proof, cupolas and shields which can be carried on tracked wagons and hauled by powerful tractors from place to place. I will enter into the nature of these defenses more fully in another lecture.

52. Protection during Battle

Protection during battle will obviously vary with circumstances. The nature of the country may be suitable for offensive or defensive action. If both sides are seeking a decision, open country, that is a good offensive area, will be sought; but generally this will not be the case, the weaker side retiring into broken country, or into a prepared anti-tank zone.

Protection during battle may be examined under three headings:

(1) Protection of the mechanized arms—the striking force.
(2) Protection of the non-mechanized arms and rear services.
(3) Protection against motor-guerilla attack.

From these three categories of protection we obtain the picture of a stronghold or castle, surrounded by a ring of outer defenses from which an armored sally party can issue forth and engage the enemy in the open field. A very similar picture to that presented to us in medieval warfare—namely, a fortified wagon laager surrounded by foraging parties and "battles" (battalions) of mounted knights in armor who challenged each other to combat outside their respective fortified posts.

The tactical base of the medieval system was the laager, in the mechanized system it will also be the laager in changed form; not a wagon fortress but an anti-tank fortress, an area of ground chosen on account of its natural strength, and protected by a chain of anti-tank cupolas and shielded guns.

[The "Knightsbridge Box" in May-June, 1942, was in fact a hastily improvised laager of the static kind.]

At a distance and around this moveable fortress will be posted the guerilla swarm, not a fixed protective outwork, but a mobile one, moving outwards when unopposed, and, when attacked by superior force, withdrawing when necessary within the mobile fortress.

As regards the protection of the mechanized force itself, being armored, its main protective power lies in its offensive action. The ground, the sun, the wind, as well as its distribution in such an order that whenever attacked it can refuse a flank and take the enemy in flank, are all protective means which must be considered in the plan of battle.

[It has been reported that General Rommel prefers to launch his tanks into attack when the sun is in the West, so that his enemy has to face its glare.]

If taken at a disadvantage, instead of risking defeat it should retire on the mobile fortress, and either reorganize itself under its protection, or seek refuge in it and refit. It may seem that this is to court destruction through siege; but such a danger will, so I think, prove the exception, for as the enemy must refuel and refit, it is unlikely that he will be able to remain outside the fortress for more than a few hours or possibly a day or two. What is more likely is that he will attempt to carry out a series of aircraft bombardments on its now crowded garrison, and simultaneously interpose himself between the fortress and its nearest advanced depot.

[Notably at Bir-Hakeim in June, 1942.]

53. Commanders in an Outpost System

I have already examined this question, pointing out that, as in mechanized warfare protection must be circular instead of linear, one commander will be required to control the protective circle instead of a number. In theory, the general-in-chief is responsible for the security of the whole of his army and its services. In present-day warfare it is easy to detail an officer to command, as the difference between tactical operations is not so great as it will be in mechanized warfare. In that type of fighting there will be a distinct difference between the tactical duties of the commander of the protective circle, of the mobile fortress and of the striking forces. Therefore, the general-in-chief will require three subordinate commanders, the command of the outposts falling to the circle commander, who, as I have already mentioned, will have under him five subordinates—the commanders of the swarm and of the four guards.

54. Miscellaneous

Under this heading fall protection by means of night movements, under cover of smoke clouds, and protection by feints.

Night offers two types of protection. The first, protection from direct observation, and the second protection through ability to move unseen from one position to another. With tanks it should not be difficult in an hour or two, to shift a considerable force several miles under cover of darkness and in consequence throw the whole of the enemy's information out of gear.

The use of smoke to create an artificial fog is well known, and its importance can scarcely be exaggerated when tank meets tank; smoke clouds are used not only to blind the enemy, but to form curtains behind which maneuvers can take place.

As mobility increases, the power of delivering feint attacks and carrying out feint movements will also increase. Protection can be gained by false concentrations, false retirements, etc., which will mislead the enemy, and cause him to commit all kinds of blunders. As this is so, the defensive-offensive is likely to play an important part in mechanized warfare, the enemy being first misled and persuaded to commit himself, and then strongly attacked when in a false position. The movement of an empty column may sometimes prove the decisive factor in a battle.

OTHER PROTECTIVE DUTIES

55. Artillery Escorts

Today artillery, which is the backbone of the attack and the defense, is generally separated from the attacker and defender by a considerable distance. Before the day of the motor vehicle, the front of an army—the deployed infantry forces—protected artillery from surprise attack. Since the advent of the motor vehicle and the tank this protection is daily becoming more and more problematical. When a guerilla swarm is used, this protection may be said to be automatic; for the guns will move and come into action within the protective circle, and, therefore, escorts of other troops will be unnecessary. Should a swarm not be used, escorts will have to be provided, these consisting mainly of anti-tank weapons which will be posted in a circle round the gun positions.

56. Protection of Convoys

On the move and at the halt the rear services will be located within the protective circle, from which will run one or more lines of communication to the base. These lines will have to be protected against guerilla attack; therefore, the farther an army moves from its base the greater will become the strain of this protection.

[As has been the case with the German lines of communications in Russia.]

The organization of a line of communications is likely to be as follows: The base depot will be established at some great center of communications, or at a port. It will be a strongly fortified area impervious to tank or motor vehicle attack and elaborately defended against aircraft. As the army advances, the area connecting it with the base depot will become the area rather than line of communications. This is likely to assume a conical form as the enemy is approached. In a civilized country this area will contain one or more railways and a number of roads, according to the supply demands of the army. One or more of these will be selected as the main line of communication, and every sixty miles or so will an advanced depot be formed. If a road is selected, and I am of opinion that for mechanized warfare roads will play a more important part than railways, the general lay-out of the communications will be as outlined in Diagram 9, **a** is the depot and **f** the field army, and **b, c, d,** and **e** are the advanced depots.

[This was true in the German campaign against France; but on account of the paucity and poverty of the Russian roads, not so in Russia.]

To carry out the protective duties the following distribution of force suggests itself: Each depot will be an extemporized fortress possessing anti-tank and anti-aircraft defenses. If spaced at 60 miles apart, each will be responsible for the patrolling of half the distance between it and the last and the next depot. This patrolling will be carried out by aircraft and motor car. Each will be allotted a force of tanks, mainly of the scout and

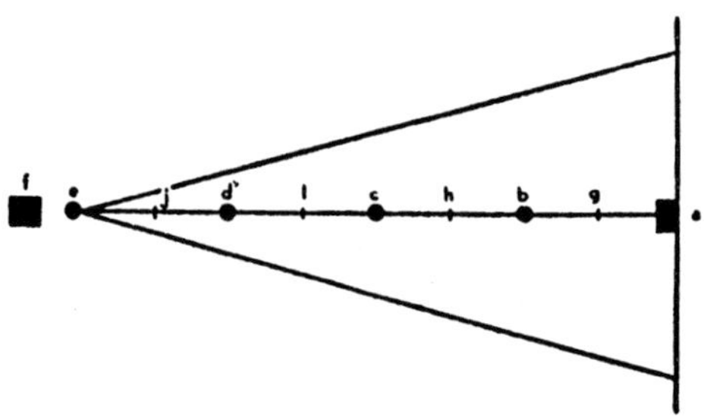

Diagram 9. The Area of Communications.

destroyer classes, which will escort the convoys to the next half-way post. Thus a convoy of vehicles setting out from the base depot will be escorted by **a** to **g,** and at **g** handed over to **b's** escort which will take it on to **h,** and so on until it reaches the field army. These escorts can work on two systems: they can either accompany the convoy, or, what is more likely, they will move out on the flanks of it and picket the margins of the area.

If a defended area of communications does not exist, then, when a convoy and its escort parks for the night, the parking place selected should be situated in an anti-tank locality, such as a loop in a river line, or in a wooded country which offers few lines of approach. In thinking out their position, defense against rear and air attack must not be overlooked.

57. Protection of Railways

The protection of railways, I think, will closely follow the system we adopted during the South African War, but with this difference: Instead of constructing a block-house line on one side of the railway, there will be a block-house line on either side, not close up to the railway but at some distance from it. The object of these two lines will be to free the area from motor vehicle attack and to prevent trains being sniped and shot at. The breadth of the area will vary with the nature of the country, but the two block-house lines should be sufficiently far out to deny an enemy observation within machine gun range of the permanent way.

58. Protection from Aircraft

As a mechanized army is comparatively small, and as its men are protected by armor, and its distribution is far more flexible than that of the horde armies of today, its need of protection against air attack will be considerably reduced, but its base and advanced depots will be as attackable as today; therefore anti-aircraft defense units should be mainly used to protect these localities. Another advantage it will possess is that such anti-aircraft units which accompany it will move in the protective circle, which itself will be able to open machine gun fire against low-flying airplanes.

As regards air protection: the most certain is obviously to gain command of the air, if not absolute then local; the next—to discover the position of the enemy, and the last—to attack him, and so as to force him to defend himself.

59. Concealment from the Air

Concealment from the air, though always a difficult problem until air supremacy has been gained, will, nevertheless, be easier in the case of mechanized armies, which in a wooded country will be able to move from cover to cover, or in open country shift their positions during hours of darkness.

LECTURE VIII

CHAPTER VII

The Attack

60. General Principles

The principles of battle do not change with the changes in the nature of weapons; what do change are the conditions of war in relation to the power of weapons. Thus, whatever weapons are used, the enemy must be found, held and hit, whereas the finder must strive his utmost not to be found, not to be held and not to be hit. Yet every weapon is influenced not only by ground, time and space, but by every other weapon—the enemy's and our own. No new weapon can be introduced without changing conditions, and every change in conditions will demand a modification in the application of the principles of war.

Once the enemy is discovered, the whole theory of the attack must be based on a careful study of ground with reference to offensive action and protective power, as well as the time it will take both sides to move over the ground. Consequently correct timing of movements becomes the decisive factor, therefore, as I have already pointed out, orders and movements must be simple, for once mechanized arms are launched their control will become difficult.

[As an example, the various British tank attack in Libya on November 18-20, 1941.]

Should ground be unfavorable, or time insufficient wherein to gain a favorable area, one side or the other is likely to refuse battle. This, I think, will frequently happen unless one side has a marked superiority in mobility, or is numerically so superior that it can accept greater risks. The idea that great battles will take place soon after the outbreak of war is not one that I hold, instead I foresee much skirmishing and maneuvering until one side commits a blunder, when, if the other recognizes it in time, a battle will take place.

[This has been disproved. In France because of lack of mechanization, and in Russia, partly, at least, on account of surprise.]

When such an opportunity presents itself, the offensive must be established on a secure base. In present-day warfare artillery provides a base for infantry, the guns decimating the enemy infantry and so easing the forward movement. In a battle between tanks, the main base is the anti-tank fortress, which is a strategical base when the tank forces are distant from it, and a tactical base when they are sheltered within it, or are operating in its immediate neighborhood.

[In 1941, many of the German tank advances in Russia, notably the first on Kiev, were no more than raids. Had they been strongly based, they would in all probability have developed into decisive operations.]

Here is presented to us an almost exact picture of the strategical and tactical uses of the medieval castle, the only difference being that the castle could not be moved from place to place. As it will seldom be possible to shift this base during battle, the duty of establishing a forward tactical base falls on the artillery tanks, whose aim is to protect the attacking machines.

[Though we possessed an efficient self-propelled 18-pdr. in 1925, we were surprised by the German 75 mm self-propelled and armoured gun in 1941.]

Though the main base can protect itself, it cannot guarantee the protection of the approaches leading towards the enemy. These must be protected by the attacking tanks, or, if this is not possible, by a special force detailed for that purpose. This means that, whenever possible, an attacking force will seek battle in an area from which, if it is unsuccessful, it can retire on to its anti-tank fortress, or into a defile—an area flanked by tank obstacles.

[In May-June, 1942, when Rommel retired after his failure to storm the "Knightsbridge Box," he fell back on an artificial defile created by his anti-tank guns and the lane he had cut through the Gazala-Bir Hkeim minefield. In October, though he occupied a defile—namely ,the El Alamein position between the Mediterranean and the Quattara Depression—the nearest defile in his rear was that of Sollum, 200 miles to the west. His position would have been considerably strengthened had he used but half the mines he actually did to cover his front, digging in the other half 40 to 50 miles in his rear. Then, were the forward line penetrated, he could have fought a delaying action along the rear line, and from thence retired to Sollum.]

As a combination of good fighting ground and good retiring ground will not be easy to find, here is another reason why battles on land, like battles at sea, will be the exception rather than the rule.

[A Special Correspondent writing in "The Times" of September 28, 1942, points out: "The use of ground (in Libya) for concealing tanks in a 'hull-down' position is no easy matter, and tanks must maneuver in view of one another in a manner of armies 100 years ago. When action is joined, there is usually little movement; the opposing forces stand face to face and swap shots until one or other has had enough . . . once a stand-up tank fight is started, each side tries to thicken up the line with field artillery and anti-tank guns, and it is often this which finally forces one or other of the tank forces to disengage."]

When such a combination of tactical values can be found, granted that the enemy is willing to accept battle, which, unless he is as fortunately situated, is granting a good deal, the next operation is to maneuver him into a position from which, in order to extricate himself, he must accept battle. This may be called the holding maneuver, for though it does not necessarily pin the enemy down to a position, it pens him up in an area from which to save himself he must get out of. The object of such a maneuver is not so much to destroy him by attack as to force him to surrender through lack of supplies. Therefore, whether the attack is to be made on his fighting-power or on his moving-power will largely depend on the nature of the area in which he is bottled-up.

The whole of these various operations will depend as much on economy of force as on numerical strength—that is to say, on using just sufficient force for each operation, so that a strong reserve may be held in hand until the enemy has fallen into a trap or committed himself to a definite course of action.

[In his mobile battles, though not in his static one of El Alamein, Rommel showed a masterly appreciation of economy of force.]

In mechanized warfare the value of a reserve force cannot be exaggerated, for increased mobility carries with it the power of effecting innumerable surprises, and the more the unexpected becomes possible the stronger must be the reserves. One of the great difficulties in the future will be to gauge the enemy's intentions, also it will frequently be most difficult to fix him in any definite locality; therefore, unless strong reserves are kept in hand, it will be impossible to meet unexpected situations.

[This has been proved again and again during the war. Reserves can be husbanded and created by constant rallying as well as by rapid field repairs. A tank out of action is seldom a total loss.]

Again, the more mobile armies become, the more difficult will it be to control them, and, unless reserves are kept in hand, control is liable to be lost, and with it command, an army rapidly being reduced to a leaderless mob.

61. Objectives and Frontages

In infantry warfare objectives are generally selected with the intention of enhancing the offensive power of one side to the detriment of the other side. In mechanized warfare their selection will more frequently be governed by the idea of enhancing and reducing mobility, for until superiority in mobility is gained offensive power is of secondary value.

Objectives must not be confused with the decisive point of attack, for objectives are like the rungs on a ladder, they are but means towards attaining a decision. The place where a decision can be gained is called the decisive point, which has always been that point in an army, generally the rear, from which its line of communications to its supply base runs. In the past it has been extremely difficult to strike at the enemy's rear without uncovering one's own line of communications; but as mechanized forces are able to move rapidly across country, it is both easier for them to change their own communications and to attack the enemy's at short notice.

[One of the characteristics of the present war is the frequency of direct tank and air attacks on lines of communication; in France, in Africa and in Russia.]

ARMORED WARFARE

Whereas today armies maneuver to protect their communications, in mechanized warfare it will frequently be necessary to maneuver communications in order to supply armies and maintain their liberty of action in the field. Because it is impossible to maneuver roads and railways, as I have already mentioned, cross-country trains are likely to take their place.

[In my opinion, lack of cross-country trains ruined the 1941 German campaign in Russia. As early as August, on account of weather conditions, tied as the supply trains were to the foundering Russian roads, the forward armoured division frequently had to retire or halt for days on end.]

When a mechanized force is used for a decisive attack, it is necessary that the enemy be first held, or that his power of mobility be reduced, so that his freedom of movement is restricted; unless this is done, rear attacks are likely to prove wasteful.

[In France, the Maginot Line automatically fixed the greater part of the French Army. In Russia the German failure to fix the Russians was the major cause of their abortive 1941 campaign. Had the money spent on the Maginot Line been allotted to the construction of medium tanks, 6000 could have been built and with them 20 armoured divisions equipped.]

In other words, the enemy must be first held, and it is from this holding operation that the rear attack should be developed. From this it will be seen that an army supplied by cross country trains working in an area of communication will be far more difficult to fix than one depending on rail or road communications; for the decisive point of attack will be flexible, it will not, so to say, occupy a position, but will move within an area.

There are objectives the gaining of which enhance protection or offensive action, also there are objectives which enhance or restrict mobility, and amongst these must be included areas and localities the occupation of which will distract the enemy's attention, and so cause him to look in more than one direction. Distracting attacks are not feints, but operations which force the enemy to change his plan or deplete his reserves.

[Propaganda is a powerful instrument of distraction, e.g. such as preceded the air attack on England in the summer of 1940. The constant cry of the German radio was "invasion," until a veritable panic was engendered.]

An objective, which will generally be an area, should never be selected without first considering its flank approaches. Will these protect an advance on it? If not, then flanking protection will have to be arranged. Again, objectives should always be related to each other, each being a step in a definite plan. Unlimited objectives, even in the pursuit, may be considered as always wrong, for they can only lead to confusion; in fact, objectives should not be far distant from each other so that forces may frequently rally and reorganize.

[In the German attack on Russia, the initial tank objectives were too distant, more especially so as the German armoured divisions lacked anti-tank wings and cross-country supply trains. The depth of an attack stands in direct ratio to facility of supply.]

From objectives I will now turn to frontages, which are not lines or distances, but areas of operations of considerable depth as well as breadth; for without depth there seldom can be security for the flanks, and without this security the actual front of attack is likely to be weakened. In mechanized warfare a frontage is not a continuous line, but rather a morcelated arrowhead, or crescent, which encloses a semi-circular area, the cord across the base of which can, if desired, be blocked by a separate force.

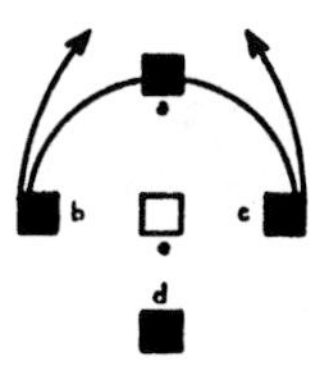

Diagram 10. Initial Frontage of Attack or Defense

If the formation assumed during the approach march, is that of an open square with the anti-tank wing in the center, then, should an attack be developed by or against any one of the four mobile groups, the frontage of attack will include the group attacked and the groups on its right and left. Thus, in Diagram 10, should group **a** attack, or be attacked, **b** and **c** at once become its offensive wings. Frontages, therefore, fall into two categories: (1) The initial frontage—that is, group **a** distributed in fighting order—and the potential frontage of **b** to **a** to **c** and the area in between, which as the battle develops will expand or contract in size. Compared to an infantry battle, there is no well defined frontage of attack, any more than there is in a naval battle. Should the action develop rapidly, then it is important to maintain the original protective approach formation until the anti-tank wing **(e)** deploys and establishes its base of action. Until this is done, group **d** should be detailed to protect it, after which it can come into general reserve.

THE GAINING OF CONTACT

62. Special Considerations

I have already dealt at some length with the gaining of contact, the success of which will depend on peace preparation, rapidity of movement and tactical judgment. Frontal attacks should, whenever possible, be avoided, and every attempt should be made to occupy such areas and localities, the holding of which will compel the enemy to change his plan. Thus will initiative be gained. Audacity founded on sound information and a psychological grasp of the enemy's higher command are the surest foundations of success.

[In 1940, the Germans weighed up the psychological state of France correctly. In 1941 they seem to have totally misread that of Russia.]

THE ATTACK

63. Special Considerations

Contact is information of the most tangible kind, an enemy met with is an enemy at grips, and, as in a wrestling match, contact is likely to be followed by much foot-play. Time still remains the decisive factor, time wherein to modify a plan according to the information contact gains. To economize time—that is, to make the most of it—depends not only on a tactical eye but on a well prepared brain. Knowledge of the characteristics of the enemy, of the personality of his higher command and the morale of his troops, as much so as the power of their weapons and machines and their tactical predilections, will speed up the assessment of information just as simplicity of idea, precision of plan and personal knowledge of the general-in-chief will speed up action.

[In their attack on Russia, the Germans would appear to have grossly underestimated the morale of the Russian soldier.]

I will repeat it again: in every plan weapons must be fitted to the ground as well as distributed according to an idea, because the battle is fought on the ground and not in the imagination. The ground, as it were, is the factor which rationalizes the idea, changing it into a reality. The method of attack is in theory always simple—namely, the establishment of a protective fulcrum upon which to move an offensive lever. In other words, a self-protective base of action from which offensive power can be launched.

In mechanized warfare encounter-engagements are likely to become more frequent, especially during the opening phases of a campaign. They will be carried out by far smaller forces than in the past, and will probably take the form of a series of advanced actions, the object of which is to create a favourable strategical situation for a deliberate attack.

Such an operation presupposes that the enemy has been thrown on the defensive, either by having been driven into an area which he did not wish to enter, or by having retired into one which he believes will assist him in defeating antagonist. As mobility increases, deliberate attacks are likely to become less frequent; but when they are made it should not be taken for granted that their object is to smash the enemy, for if he can be starved into surrender the end will be even more certain and the means far less costly.

[This, I think, is correct, see the first annotation on Page 96.]

Starvation does not so much mean lack of food as lack of

gasoline. To compel an enemy to keep on the move, to sever him from his supply organization, and then to drive him into an area from which he cannot escape, not necessarily because it is enclosed by obstacles, but because his gasoline supply is not sufficient for him to do so, are likely to prove more effective methods of waging battle than a head-on attack.

To assist in this type of warfare, warfare against motive-power rather than against man-power, mosquito attacks waged by the motor-guerilla forces against the enemy's area of communications are likely to prove very effective.

From the deliberate battle I will turn to three tactical operations which a mechanized army will constantly have either to face, or avoid—attacks on villages, woods and river lines.

A village is not a suitable tank objective, because armoured machines are largely restricted to the roads and can be put out of action by bombs thrown from the upper storeys of the houses.

[This has consistently proved true, and more especially so in the case of large cities, such as Leningrad and Stalingrad. There is no anti-tank obstacle so formidable as a great town. The most effective city-taker is gas.]

In these operations tanks should be used to work round the flanks of the village and cut its communications. If tanks are used frontally against the village itself, they should move under the protection of the artillery and destroyer machines, ready at a moment's notice to open fire on hostile anti-tank weapons. Meanwhile, anti-tank machine gun barrages should be placed on all positions where anti-tank weapons are likely to be located.

Woods are not suitable objectives for the larger types of tanks, but if they are clear of undergrowth and brushwood, scout tanks should generally be able to move through them. When, however, these machines are unable to traverse a wood except by the roads, tracks and rides, riflemen should precede the machine under cover if the wood on each flank of the track, ready to open fire on any anti-tank weapon which may block the way. The operation here suggested is similar to that of an advance in bush warfare.

[See "Lectures on F. S. R. II, p. 154; and Appendix I.]

Again, tanks should, when possible, operate against the flanks and rear of the wood.

Excepting amphibian machines, an unfordable river is a definite tank obstacle. When a river has to be crossed, the first problem is to form a small anti-tank bridgehead, the next to raft a number of scout tanks over; these, on landing, should at once increase the size of the bridgehead by operating outwardly against the enemy holding the bank. Under cover of this protection, bridges will then have to be built for the heavier machines. Though a river is a definite tank obstacle, it should not be overlooked that, as tanks possess a high mobility, many more crossing places can be threatened than could be by means of infantry. Feints are, therefore, likely to play an important part in future river attacks.

64. General Conduct of the Attack

The conduct of the attack will first depend on the correctness of the distribution of arms to areas and objectives, and to the forward movement being hinged on a strong protective anti-tank base: secondly, on the ability to restrict the enemy's mobility by denying him certain areas and compelling him to move into others, and thirdly, upon maintaining control over the various forces until victory is gained.

As regards the second, artillery tanks will play an important part, for by using smoke shell they are able to blind the enemy's O.P.'s and by bombarding areas and approaches will render maneuvering in and through them hazardous. Artillery barrages, as employed during the World War, will seldom be used, unless frontal attacks are unavoidable, or the mouth of a defile has to be blocked, but machine gun barrages on anti-tank positions are likely to be frequently resorted to as a protective measure. Normally the attack is likely to progress through the following stages:

(1) Movement forward from the anti-tank base.

(2) Maneuvering for position, and distracting attacks to force the enemy to change his plan and draw on his reserves.

(3) Driving the enemy into a corner from which he will have to

fight under a disadvantage, or succumb to gasoline starvation.

(4) Movement forward of the anti-tank base and the handing over of the conquered area to the army of occupation.

Of the many problems which will occur during battle, probably the most difficult one to solve will be the maintenance of control.

[In Libya this has, again and again, been proved correct, and no doubt in Russia also.]

Radio-telephony will undoubtedly assist, nevertheless jamming is likely to be frequent; consequently this form of communication cannot be solely relied upon. I will repeat it again; to control tanks in the attack, the plan of operations must be as simple as possible, and it must be known in detail to all leaders taking part. Advances should be made by map squares, units frequently rallying and reporting their positions. Touch between neighbouring units must be maintained and air information must be constant.

To complicate and weaken the enemy's control, his headquarters—stationary or mobile—once located, should be bombed, and whenever an opportunity offers, the guerilla swarm should penetrate into his area of communications to distract his attention and attack his nerves.

LECTURE IX

CHAPTER VII (Continued)

The Attack

65. Infantry in the Attack

In Lecture II I examined the types of infantry soldiers which can be profitably employed in mechanized warfare. I pointed out that infantry, as known to-day, are out of place in the mechanized battle, and that what we now require are:

(1) Field pioneers armed with anti-tank weapons and transported in cross country vehicles, to establish the defenses of the army of occupation.

(2) Field police armed with machine guns, rifles, and possibly also non-lethal gas appliances to occupy and organize the conquered areas and territories.

(3) Light infantry armed with light machine guns and rifles to operate in areas unsuited to tank movement—such as forests and mountains.

I have already dealt at some length with the first, and, therefore, in this section I will consider the third.

To-day infantry are still trained for the assault. In other words, the whole of their fire tactics, machine gun and rifle, pivot on the bayonet.

[Romantically the bayonet assault is immortal. Time and again during this war Correspondents have sung its praises. Yet, so long ago as 1865, Surgeon-Major G. Hart wrote of the American Civil War, that he saw few wounds "except accidental ones . . . I think half-a-dozen would include all the wounds of this nature that I ever dressed."]

In open field land the assault expired over seventy years ago, and as a leading tactical idea it is obviously inapplicable to forest and mountain warfare, because an assault demands a close formation, and forest and mountain warfare require an open one, a series of individual groups and not a collective human wall.

The secret in mountain and forest warfare is identical to the secret in mechanized warfare—namely, the maintenance and protection of mobility. As tank formations should have their flanks and rear secured, so should light infantry formations, and, speaking generally, light infantry cannot do better than adopt the arrow-head and open square orders. In a tank force the scout tanks reconnoitre for the combat machines, in the light infantry force the riflemen reconnoitre for the machine gunners. They move forward, search the ground, select positions and then, as the machine gunners come up, move to the flanks, the machine gunners fixing the enemy with their fire. Under cover of this fire the riflemen should once again move forward and be followed by the machine gunners, until the enemy is hedged round by a circle of bullets and is definitely pinned down. When this fixing is accomplished, the final act is not to assault him, but to interpose a force between him and his line of retreat and so compel him to surrender.

It will be seen from this that there are marked resemblances between the tactics of light infantry and tank tactics. The central idea in both is to maintain mobility by protecting it and by striking at the enemy's mobility. So long as the assault was a profitable operation, frontal attacks were not altogether useless, but right through the history of war assaults have always proved costly, and never more so than to-day on account of ever-increasing bullet power.

The frontal threat and the frontal holding attack are quite different operations. The object of the first is to compel an enemy to assume the defensive, and of the second to force him to maintain it; in other words to pin him to a locality. Once this is accomplished the true attack takes the form of a flank or rear maneuver.

66. Artillery in the Attack

The future of artillery in the attack bears little or no relationship to our existing use of this arm, because the difference between artillery in mechanized and present-day warfare is the difference between mobile and static fighting. By this I do not mean that static operations will not take place, they certainly will, as I have shown in the case of the anti-tank base of a tank attack; but normally they will be divorced from the mobile operations, and will not generally be able to assist them by artillery fire. The position we arrive at is very similar to that found in naval warfare, for if the tank is compared to a ship, then its anti-tank base may be compared to a defended port or harbour; but with this difference—that it is mobile, it can be packed up and moved from place to place, whereas the harbour cannot be.

The present idea underlying all artillery tactics is to assist the infantry to approach, to attack and finally to assault. It is in fact the butt end of the tactical fishing rod, the top joint of which is the bayonet. The bayonet has long now been rendered obsolete by the bullet, and to-day the bullet is being more and more restricted by armour; therefore we are left with the shell as the principal weapon, the shell which can pierce armour and so put the machine out of action. Once it is out of action, to all intents and purposes the soldier is unarmed and unarmoured.

As this is undoubtedly what is taking place in military evolution, the logical result is that battles are becoming artillery duels instead of bullet engagements. What does this mean? It means that the idea of a special artillery tank is really a misnomer, for all tanks are artillery—mobile armoured gun mountings, and the main differences between them are the types of guns they mount. Guns, as I have shown, will fall into two tactical categories—those for in-fighting and those for out-fighting.

[In northern Africa, until the summer of 1942 the out-fighting power of British tanks was inferior to the German. This was remedied by the arrival of the American Grant, Lee and Sherman models.]

When we turn from these considerations to artillery in the attack, it will at once be seen what we really mean is tanks in the attack, for the guns of the anti-tank base will not normally take part in the battle any more than coastal fortress guns normally take part in a naval battle; for though these anti-tank base guns will most certainly be motorized, if not mechanized, their object is to protect and not conquer an area. In siege warfare it may well be otherwise, for then the anti-tank base will be moved close up to the position to be attacked, and will, as in present-day fighting, cover the assault.

From what I have now said it will be seen that to fit our present conception of artillery tactics into mechanized warfare is absurd. Command in the artillery attack to-day is centralized in mechanized war it must be de-centralized, as decentralized as is rifle fire in a present-day battle. Once, and not so very long ago, the musket fire of a battalion was as centralized as artillery fire now is. The introduction of the rifle decentralized it, so also will the introduction of mobile armour decentralize artillery fire and with it artillery command.

67. Tanks in the Attack

The problem of tanks in the attack is by no means an easy one to discuss, as no practical experience exists whereon to base discussion. In the last war tanks were so closely tied down to infantry that little can be learnt from the battles of that period. Since the war the tank idea has been so coufounded by a jumble of other ideas, such as artillery covering fire, anti-tank protection, co-operation with infantry and even with cavalry, that it has become hopelessly entangled.

To cut this Gordian knot, it would appear that the only thing to do is first of all to sketch out as simply as possible the various actions which are likely to precede engagement, and then to take an exceedingly simple plan of battle and from it work out what is likely to happen.

The first notification of the enemy's approach will be signalled by airplane, and actual contact will almost certainly take place between his and his opponent's armoured cars and guerilla swarms. If X be the enemy and Y his opponent, what action will follow?

It may be taken for granted that X will not blindly move forward, that he will decide on no definite action until information is sufficient to justify it.

[This would appear to have been Rommel's method in North Africa.]

Meanwhile armoured car and guerilla skirmishes will develop, and possibly also feints and distracting movements.

If both sides are determined to fight it out, which will seldom be the case, both will have to decide on a plan of attack, in which the controlling factors will be: the strength of the enemy, the nature of the ground and the position of his anti-tank base, which may be near or distant. Should the two sides be fairly equally matched, then X's object will be either:

(1) To annihilate Y's tank force and besiege his anti-tank base.

(2) Or to cut Y's tank force off from its base, and compel it to surrender through gasoline-starvation.

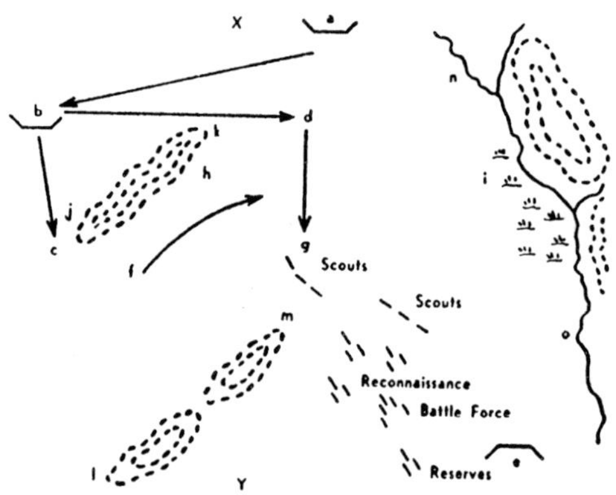

Diagram 11. Example of A Tank Battle.

(3) Or to compel Y to evacuate the area he is operating in, and retire to some other area.

To take a simple example (see Diagram 11) X and Y are in the area shown. X decides to move his anti-tank base from **a** to **b,** and make a feint towards **c,** in order to draw Y in that direction. Next, to move under cover of darkness to **d,** and from there interpose himself between Y's tank force and its anti-tank base at **e,** his object being the second of the above alternatives—namely, to force his enemy to surrender through lack of gasoline.

Y, through his guerillas, obtains information of X's change of base, and, equating this change with the topography of the area, he becomes suspicious that X's movement towards **c** is a feint; therefore, instead of remaining in **f** area, he leaves a skeleton force there and withdraws his main force back towards **e.**

X, unaware that this movement has taken place, moves early on the following day from **d** towards **g;** but Y having guessed his intention has decided to move out from his position near **e,** and by advancing on **h** turn X's right flank, and then drive him towards the marshy and hilly country at **i.** To do so he distributes his force as follows: A force of reconnaissance tanks, protected on their right flank and rear by scout machines, to advance on **h;** the detachment at **f** to co-operate on the left flank of this advanced guard, the object of the **f** force being to engage the enemy and induce him to turn towards it. The main battle force to move under cover of this advanced guard protected by scouts on its right forward flank, and in rear of this force the reserves. The object of the battle force is to strike at X's left flank and rear when he turns to engage the advanced guard.

X, however, does not fall into this trap. On meeting Y's advanced guard he is not certain whether it is moving from **f** with the main army in rear of it—that is, whether Y is operating in the area **j k l m**—or from **e** into area **k n o m.** Learning through his air scouts that his enemy's forces are in the second of these two areas, and not wishing to engage him frontally, he slowly falls back, and eventually retires to his anti-tank stronghold at **b.** Though a certain amount of skirmishing will have taken place, no battle is fought, because neither side has compromised itself, such a conclusion I consider will frequently occur in mechanized warfare.

[Much of the tank fighting in North Africa has borne this out.]

From this general outline of the battle I will turn to minor tactics and examine two operations in detail—namely, the action to be adopted against a moving enemy and against a stationary enemy.

(1) **The problem of tank against tank.** As this problem is one concerning which we have no practical war experience worth talking about, my suggestions are purely hypothetical. It would appear that, normally, a melee will take place, in which it will be difficult to distinguish friend from foe.

ARMORED WARFARE

[This was most marked between November 18-25, 1941, when General Cunningham attacked General Rommel south of Tobruk.]

The whole problem is really one of maintaining control. If the scout tanks move ahead of the combat machines, it is obvious that greater order will be maintained, as the formation of the combat machines is not so likely to be broken up. A time, however, will come when scout will engage scout, which means that this protective screen will rapidly develop into a mobile firing line. What, then, should be the general tactical theory upon which this firing line and the combat or main battle force should work? The answer to this question is, I think, to be found in the general configuration of the ground. The ground is unlikely to be dead level, it probably will be accidented in places, and so will offer covered approaches. The utmost use must be made of these in order to protect movement, and above all to take up positions from which fire can be opened **at the halt.** In other words the art of attacking will largely consist in establishing moveable strong points from which carefully directed fire can be brought to bear on the enemy's machines, whilst other forces of moving machines drive him towards them.

[The Germans have frequently used tanks in this way, both in North Africa and Russia. At bottom it should never be forgotten that a tank is no more than a self-moving, armoured field piece.]

Given a force of scout, combat and artillery tanks, Diagram 12 illustrates these tactics in simple form, and they are reminiscent of the old tactics of horse artillery working with cavalry. The enemy's line of advance is shown by the letter **a,** and our own forces are distributed as follows: **b^1, b^2** and **b^3** are the scout, **c** the combat and **d** the artillery machines. As the enemy advances he is engaged by **b^1,** under the cover of whose fire **c** comes into action, and a running engagement ensues. Meanwhile, under cover of the ground **d,** protected by **b^3,** moves rapidly to **e,** towards which **c** attempts to drive the enemy. If **c** succeeds in doing so, then directly the enemy comes within effective range of **e,** fire will be opened on him by **d.**

Though it seems to me it will be the exception rather than the rule that these tactics will result in startling or rapid decisions, the more often this simple maneuver of hammer and anvil is repeated, the greater will be the losses of the enemy, and as his offensive power is reduced, he will be compelled to retire from area to area until eventually he is cornered and starved out.

Diagram 12. The Hammer and Anvil Maneuver.

Though this example of tank battle tactics is simplicity itself, to show how complicated it may become, all that is necessary is to introduce on both sides half a dozen such forces as I have dealt with. Then it will be seen that, in order to concentrate effort and to prevent dispersion, control must of necessity become an exceedingly difficult problem; for there will be, according to plan, a general relationship between the six groups, or squadrons, used by each side, as well as an individual relationship between any one group and its neighboring groups, as influenced not by plan but by the enemy's action. In order to command one group against a single other group, command of the air will be of great assistance; but to command one group against several groups, or several groups against several groups, command of the air becomes an essential, for without it, even if the battlefield be exceptionally open, it will be next to impossible to discover the various positions and movements of the enemy.

In these battle tactics there is one minor point which requires examination, and that is—how will tank engage tank? I think the eventual answer to this question is likely to be that normally tank will not engage tank, but instead that tank unit will engage tank unit. Of course should a single tank meet a single tank a duel between them will take place; but if control is efficient, such engagements are likely to prove the exception rather than the rule, for a multiplicity of such engagements will show that control has been lost.

Again, when a tank unit engages a tank unit, fire should not be opened in a haphazard way, but should be concentrated first on the enemy's command tank, if this can be distinguished, or on his leading machine, and, should this machine be put out of action, then on the next to it.

(2) **The problem of tank against anti-tank weapons.** Omitting land mines, the two true anti-tank weapons are the gun and the armor-piercing machine gun, both can be used from within or without a tank. When used without, though they are deprived of their mobility, they gain immensely in accuracy of fire; for there can be no comparison between the accuracy of a gun fired from a stationary platform with one fired from a rapidly moving one.

[It is this characteristic which differentiates anti-tank fire from tank fire. Though the anti-tank gun proper should be transported in a cross-country machine, its most effective firing position is on the ground. Being comparatively small it is inconspicuous and can generally be concealed.]

When a moving tank is compelled to attack an unarmored gun in the open, the most dangerous range lies between 1,000 and 800 yards.

[This refers to conditions as they were in 1932. Whether they have since been modified, I cannot say. A 25-pounder firing from a stationary position should be effective up to 1500 yards, possibly more.]

Though at closer range the tank will offer a larger target, the nearer it approaches the more deadly becomes its own fire. Further still, if it advances diagonally—that is, at an angle to the gun's line of fire—the nearer it approaches the gun the more rapidly will the gun have to be traversed, and the more rapidly it is traversed the more inaccurate becomes its fire. Should this happen within 500 yards of the gun, in nine cases out of ten it will be put out of action.

When, however, we consider the armored gun, whether in a tank or in a bullet-proof cupola, the problem of attack is far more difficult; for though the disadvantage inherent in traversing on a moving target remains, the fact that the tank's machine guns are neutralized by the gun's armor will reverse the odds. The nearer the tank approaches, whether obliquely to the gun's line of fire or not, the more likely it is that it will be hit. The problem which faces us here is very similar to that of a warship attacking a coastal fortress.

This problem of the attack is further complicated in the case of the armored anti-tank machine gun, because as this weapon is comparatively small, it can be hidden away, and can open surprise volume fire at close ranges, ranges at which it could scarcely miss an approaching tank.

[This weapon is to all intents and purposes obsolete.]

From this it will be seen that the frontal attack should be avoided unless adequate preparations are made to cover it. As these demand time, and as field attacks are rapid, time will seldom be available to arrange them.

The general conclusion is that armored stationary anti-tank weapons if well sired are unprofitable targets for moving tanks to attack.

[This has been proved again and again in North Africa.]

Certainly if the attack is a frontal one, and even if it is made in flank, or in rear, it must be remembered that a gun or machine gun in a tank, or armored cupola, can be switched round in a few seconds to meet such attacks. The more mobile the defense is—that is, the more able it is to shift its weapons in a defensive locality—the more powerful it becomes; therefore, the most formidable anti-tank weapon is the tank itself—a weapon which can move, is armored, and can fire either when moving or when standing still.

68. Cavalry in the Attack

Though cavalry can play no part in the battle between armored machines, it should never be forgotten that the cavalry idea is essential to mechanized warfare. And it is because armor enables the mobility formerly expressed by cavalry, once again to dominate the battlefield, that mechanized warfare is superior to infantry warfare.

In the days when cavalry were supreme, horsemen could often operate as a separate arm; yet, as an art, cavalry tactics reached its zenith when mounted troops were combined with foot soldiers, because foot soldiers were able to establish a protective tactical base from which cavalry could develop their mobility and operate offensively. Examples of this are to be found in classical warfare, such as the battle of Arbela (B.C. 331); in medieval warfare, such as the battle of Dupplin Muir (1332), and more recently in the campaigns of Frederick the Great. After the age of that great soldier cavalry began to decline in value, and after the introduction of the rifle they were rapidly divorced from their infantry base, because their mounted offensive power was increasingly reduced by the bullet.

The static nature of the World War was due to the fact that no sustained mobility could be developed from the infantry base. It was a war of protective tactical bases; therefore trench or siege warfare predominated, every attempt being made, now that it was no longer possible to develop cavalry mobility from infantry stability, to develop infantry mobility from artillery stability. This change in tactical method, however, failed, for it was no more possible for infantry to advance against the bullet power of the modern rifle and machine gun than it was for cavalry to advance against the musket and case shot firing cannon of three-quarters of a century ago. What was required in order to solve this problem was a bullet-proof horse and horseman, and the answer to his problem was the tank.

It was because cavalry tactics were, right through history, at their best when cavalry mobility was linked to infantry protective power, that I have made such a point of combining tank forces with anti-tank forces, and tank action with anti-tank action, looking upon the one as the battle hammer and upon the other as the battle anvil; in other words—complementary instruments. If ever there should be a supporter of the tank idea it is the cavalry soldier, for the tank idea of today is the cavalry idea of yesterday.

69. Engineers in the Attack

In the mechanized battle, in every sense of the word, the engineer is a combatant, and one of the most important elements in the anti-tank organization.

[This is one of the outstanding lessons of this war.]

It is for him to design the anti-tank defenses—what works are to be dug, what minefields laid, what bridges constructed or demolished, what gas inundations employed and what anti-gas defenses established. To assist him in these various duties, future wars will, in all probability, see him equipped with anti-tank trench diggers, mine layers, mine sweepers or exploders, and bridging machines, as well as tanks equipped to emit clouds of gas and toxic smoke and drench areas with vesicant chemicals.

[Mine sweepers have been used in this war, but apparently without much success. Had General Montgomery possessed an adequate number of good mine sweepers on October 23, 1942 he could have penetrated General Rommel's El Alamein minefield in twelve hours, instead of twelve days, and it would have been unnecessary for him to have sent infantry forward to grub up the mines. In 1918 bridging and gas producing tanks, as well as other special machines, were either built or building; yet, so far as I am aware, no belligerent has made use of them in the present war. Inventive genius would seem to have concentrated purely on offensive types. In my opinion this is wrong, because special and often non-fighting machines are essential in gaining surprise.]

70. Aircraft in the Attack

The object of the airplane, when combined with a mechanized army in the attack, is a threefold one:

(1) To gain local command of the air either by directly defeating the enemy's air force, or by drawing it away from the battlefield by attacking some distant centre the defense of which is vital.

(2) To gain information by locating the enemy, maintaining contact with him, watching his movements and rapidly reporting them.

(3) To protect the attacking troops by carefully reconnoitering the area, and all localities in which hostile guns and anti-tank weapons are likely to be found. Unless the enemy and the ground are carefully watched by aircraft, tanks will be denied much of their power in the attack.

The tactical base of air action is not some far distant aerodrome, but the anti-tank base, because, in order to save time and make the most of information, there must be the closest possible co-operation between the mechanized arm in the air and the mechanized arms on the ground.

[The loss of the island of Crete was largely due to lack of this co-operation.]

The anti-tank base should, consequently, include an air landing ground and air defense troops; for, as it is likely to be attacked by enemy aircraft, without such troops it cannot form a protective base for aircraft.

From this we see that if the entire forces are to be efficiently articulated, not only must the air force be closely combined with the air defense force, but that both these forces and the field army should come definitely under one command—**Tria juncta in uno.**

LECTURE X

CHAPTER VII (Continued)

The Attack

FURTHER CONSIDERATIONS FOR THE ATTACK ON AN ORGANIZED TRENCH SYSTEM

71. Special Conditions

The special conditions of yesterday frequently become the general conditions of tomorrow. When firearms were first introduced they were special weapons; yet in a century or two they became general, entirely changing the form of war. Before their introduction warfare had largely consisted in forays from strongholds. Knights sallied out from their castles, battled in the open field lands and then retired to them, and the only certain process of reducing these bases of operation was to besiege them and starve them out, for their walls were generally unshatterable by the projectile weapons of that period. Then came the cannon which could overthrow rampart and bastion, and by doing so it established the offensive as the stronger form of war. Lastly came the rifle, which, on account of its superiority of range and accuracy of fire, so widened out the distance between contending armies that the assault, and eventually the attack, became less and less possible. Therefore, as neither side could leave the field without proclaiming its defeat, in order to maintain its position it dug itself in; in other words, it built for itself a bullet-proof castle and went into bullet-proof quarters.

The World War conclusively showed that, though with the weapons which existed in 1914 it was possible to carry out forays from these castles—entrenched lines—it was most difficult to shatter them, even by the most intense concentrations of artillery fire. The result was that the war took on a pronounced siege form, the object of each side being to starve out the other; in fact, to revert to twelfth and thirteenth century warfare. Once again the defense had definitely become the stronger form of war.

What the final stages of this war taught us was that without gasoline there can be no true mobility when horde armies face each other. Even if unarmored a comparatively small motorized force should be able to worry to death a horde, just as a flight of sparrows can worry to death a hawk. It is for this reason that, earlier in these lectures, I have made so strong a point of the motor-guerilla, and though he is mainly restricted to road movement, his mobility is so superior to the foot soldier's that he should generally be able to throw him on the defensive.

The fundamental strength of the motorized soldier is that he can surprise; his fundamental weakness is that he cannot close—he can worry his adversary to death, but he cannot strangle him. Encase him in armor and this weakness vanishes; endow his vehicle with cross-country movement and his "locomobility" is vastly increased. Thus the motor car and the tank re-establish the offensive as the stronger form of war. Nevertheless, this fact does not justify us in supposing that mobile warfare will endure for ever; but rather that sooner or later, means of slowing down tank offensives will be resorted to, and once again the castle, in modified form, will appear.

[This has been noticeable in Russia; for, as the war proceeds, more and more do defensive works play a dominant part. The result is an ever-increasing demand for infantry—static troops.]

Once again armies will be faced by siege warfare, but in what forms?

I think the answer to this question is that there will be two—the linear defense and area defense; the first being resorted to so long as the horde idea lasts, and the second gradually replacing it as the tank idea grows, or, perhaps more accurately expressed—the anti-tank idea. The first will consist of trenches as we knew them in the World War, reinforced by anti-tank weapons and devices; and the second, of a zone of mutually supporting anti-tank works, which will not only form a shield against tank invasion, but by protecting all forces in rear of it will establish a secure air base from which air attacks can be launched on the enemy and his people.

72. Issue of Orders, Conferences and Preliminary Arrangements Prior to an Attack on an Organized Trench System

The arrangements which fall under this Section will depend on whether anti-tank defenses have been constructed during peace time or war time, and also upon whether both sides are stalemated.

Should anti-tank zones be constructed during peace time, which, on account of cost is unlikely, then the "breaking" of them will, in all probability, be handed over to the respective air forces.

Should, however, the defenses be constructed during war time, the first to be built will undoubtedly, be those which block the main lines of approach.

[This may be observed in the general set-out of the 1941-1942 German winter line in Russia. Rail centres, such as Rzhev, Smolensk, Bryansk, Orel, Kursk, and Kharkov are firmly held. The curtain walls in-between are garrisoned by little more than observation troops.]

As this throttling of mobility will detrimentally affect offensive action, it may be expected that battles will be fought with the object of preventing their construction. We are thus faced by two forms of siege warfare; the object of the first is to penetrate a peace built zone, and of the second to frustrate the building of a less formidable zone during war time. In other words, attacks on permanent fortifications and attacks on improvised fortifications.

In the first case, offensive preparation will obviously be made during peace time, a number of plans being worked out in detail according to the number of sectors which can profitably be attacked. In the second, plans of attack will have to be improvised at short notice, and, so it seems to me, will mainly be based upon threatening the enemy in one direction and hitting him in another. Though time wherein to prepare these attacks will be longer than for attacks in field warfare, it will seldom if ever be sufficient to permit of the ritualistic methods of mounting trench battles as witnessed in the last war.

73. Special Preliminary Measures in the Attack on an Organized Trench System

In the trench battles of the World War, up to the date of the battle of Cambrai (November 20, 1917), method ousted surprise and brute force cunning. In attacks on anti-tank belts and zones of works, method must go hand in hand with surprise, and cunning must reinforce brute force. Nevertheless, in all these operations of penetration, as the main object is to break through a series of fortifications, whether connected or mutually supporting, mobility must be subordinated to offensive power—that is to say, works must be silenced, or destroyed, before a maneuver can be contemplated. To accomplish this, two weapons will be required: The high explosive shell and the gas bomb, shell, or gas inundation. This necessitates the use of a siege train.

[The siege-train should include mine sweepers and gas tanks, as already noted, mustard gas is the ideal city taker.]

The more mobile this siege train is, the more rapidly will surprises be effected, it must therefore be able to move across country.

74. General Conduct of the Attack on an Organized Trench System

As I have already pointed out that in siege warfare, the form of the tank attacks will depend on whether the enemy's defensive system consist of a belt or a zone of works, I will now consider these in turn.

(1) **Attack on a belt of works.** For an attack on a belt of works, that is, on several lines of entrenchments—the main obstacles which the attacking tanks will meet with are those which will impede mobility—namely, trenches and the anti-tank weapons.

[The Maginot Line was of this description.]

If trenches cannot be spanned, obviously the attack will fail, and if the anti-tank weapons cannot be silenced then the attacker is certain to lose heavily.

At the battle of Cambrai, the first difficulty was overcome by using tank-fascines, large bundles of ordinary fascines which were tipped off the roof of a tank and pitched into the trench in order to support the tail of the machine as it crossed over. Later on in the war, these fascines, which were very heavy, were replaced by hexagonal cribs. Both these appliances enabled a 26 foot tank to cross a 15 foot gap. The second difficulty was solved by the use of smoke shells.

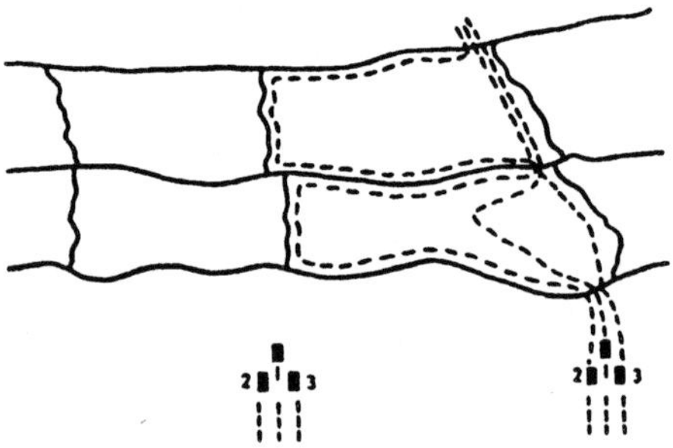

Diagram 13. The Cambrai Attack.

In battles of this nature, the difficulty is not to cross a single trench line, but a number of lines, and as each tank can carry only one fascine or crib, a rather complicated maneuver has to be resorted to. At the battle of Cambrai it was as follows:

The entire front to be attacked was divided into a number of tank areas, each area being bounded by trenches—that is to say, it had a trench in front and a trench in rear of it, and on its flanks ran communication trenches. As there were three lines of trenches to be crossed, to each area was allotted a section of three tanks, each carrying a fascine. In rear of each section came three separate parties of infantry: (1) Trench stops which halted at the point crossed by the leading tank, and there blocked the trench. (2) A trench clearing party which, under protection of a tank, drove the enemy down the trench towards the next stop on the left. (3) A party to garrison the trench whilst it was being cleared and to act as a reserve.

How the tanks worked is shown in Diagram 13. The leading tank, without infantry following it, moved forward, crushed a path through the wire, dropped its fascine, crossed the trench and moved into the center of the tank area where it become an advanced guard. The two following tanks, Nos. 2 and 3, crossed over No. 1's fascine, No. 3 with its infantry moving straight ahead, whilst No. 2 and its infantry, less the trench stopping party, moved to the left. At the second trench No. 3 dropped its fascine, as No. 1 had done at the first, acted in the same way as No. 2, and Nos. 1 and 2 then followed it. At the third trench No. 2 dropped its fascine, etc.

It will be seen that the tactics of this attack were exceedingly methodical; nevertheless, the attack came as a surprise.

Before I attempt to modernize these tactics, it will, I think, be of some interest briefly to outline two other plans of attack which were considered during the war.

The aim of the first was to cut off the garrison of an entrenched area from the reserves in rear of it by carrying forward machine gunners in tanks and depositing them in rear of the entrenchments. Large machines were actually built for this purpose, each of which could carry four machine gun teams of five men each, besides the tank's crew. They were not a success, not because the idea was unsound, but because they were underengined and far too slow.

The second was to pass a fleet of long high-powered machines right through the entrenched belt and attack the enemy's headquarters in rear of it, whilst his front was being attacked by tanks and infantry on the Cambrai system. The idea behind this form of attack was to paralyze the enemy's command, and, therefore, render his army inarticulate. It was novel, but the war did not last long enough for it to be tried out.

[For full description see my "Memoirs of an Unconventional Soldier," Chapter XIII.]

Bearing in mind these three methods, the Cambrai tank and infantry attack, the carrying forward of machine gunners and the attack on the enemy's command, I will now attempt to describe to you, how, with the machines considered in these lectures, these three forms may be amalgamated and the whole attack improved.

The first thing which strikes us is that, with machines of increased mobility, we should be able vastly to speed up the attack. This undoubtedly is so, and every advantage should be taken of it. The second, which is not so apparent, is that if tanks can be divorced from infantry, the total frontage of attack can be doubled, perhaps trebled. The reason for this is that without infantry it is not necessary to advance the tanks in the close order adopted at Cambrai, but instead in an open order—that is, with a considerable interval between each group of tanks and between the objectives of each group. This perhaps needs a little explanation.

[See Appendix I.]

If you turn to Lecture X in "Lectures on F. S. R. II" you will find a full description of the difference between using infantry or tanks in an attack of penetration; the main one is, that when infantry attack, the flanks of the advance are pushed inwards, whereas when tanks attack they can be pushed outwards, as shown in Diagram 8 of the lecture quoted. At the battle of Cambrai the front of attack was continuous, so that the "wall" of moving tanks would protect the infantry following each

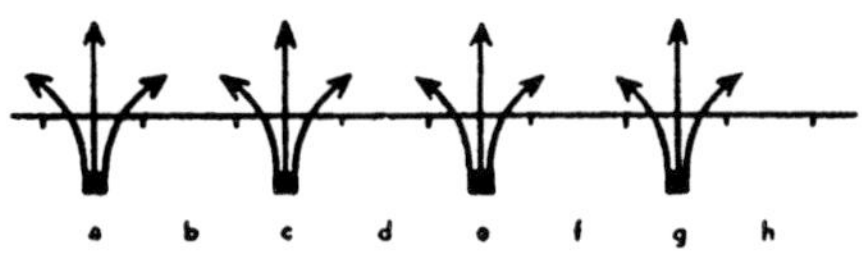

Diagram 14. Attack on A Morcelated Front.

machine. Had it been morcelated, as shown in Diagram 14, it would have been possible for the enemy in the sections not attacked—namely, **b, d, f** and **h** to have brought enfilade fire to bear on the infantry, which might have resulted in the failure of the plan.

[In June, 1941, the German tank front in Russia was a morcelated one, and in spite of the vast stretches of country separating their armoured attacks. These were, generally, carried out by two separated armoured groups, which, as they advanced, left the Russian forces in-between them in a salient. For a description of these tactics see my "Machine Warfare,' Part II, Chapter X.]

Accepting a mórcelated front, and granted that there are three main trench lines to cross, and that these trenches are as broad and deep as they were at Cambrai, the general distribution of tanks should, I suggest, be as follows:

First Wave. The object of the first wave is to penetrate the whole entrenched belt in the least possible time, and to pre-

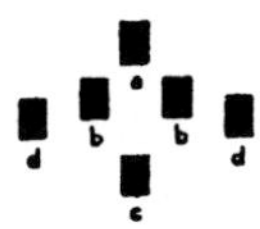

Diagram 15. Assault Formations.

pare the advance of the second and third waves.

[For example, the initial German tank attack in the vicinity of Sedan in May, 1940.]

It should consist of assault **(a)**, destroyer **(b)**, bridging **(c)** and smoke producing machines **(d)** formed up as shown in Diagram 15.

[Also mine-sweepers, should minefields have to be crossed.]

The assault tanks will move straight forward crossing the trench with or without cribs, according to their width. Following them will be the bridging machines which will bridge the trenches so that the smaller tanks can cross them.

[In 1932 I invented an automatic tank bridge which, as it formed part of the tank, enabled it to cross any number of trenches in succession up to 25 feet in span. Needless to say, it was not adopted. Had the German light tanks been fitted with this bridge, they could, with ease, have negotiated the dykes of Dunkirk. The whole question of tank bridging has been shockingly neglected.]

Over these bridges the destroyer tanks will pass, their duty being to work out on the flanks of the assault tanks and protect them from anti-tank fire. As regards the smoke producing machines, they will have to work according to the wind, and the most favorable direction is when it is blowing towards the enemy, the next towards the attackers, and the least favorable when blowing from a flank—this is, at right angles to the advance. In the first two cases, these machines should move out on the flanks of the assault, generating a wall of smoke on each side of it. In the last, in advance of the assault, forming a wall of smoke in front of it. At the bridges all machines will have to close inwards, but once over, the destroyer tanks and the smoke producing machines should move outwards.

Second Wave. The object of the second wave is to clear the trenches, protect the first wave from counter-attack and establish a forward anti-tank base. It should consist of combat and scout tanks, anti-tank troops and possibly also infantry in armored carriers. The combat tanks should, obviously, work in large groups, two or three to the whole frontage attacked, and not necessarily a group to each sector attacked.

[The German method in France and Russia.]

If in three groups, then probably one will move on each flank of the frontage of the attack and one in its center. Under their protection, as well as that afforded by the first wave, an anti-tank base should next be established by the anti-tank troops forward of the area penetrated, which will form a rallying-point as well as a base of operations and a harbor of refuge should the enemy bring into the field a superior tank force.

[As I have noted, the lack of such a base ruined many of the German tank advances in Russia.]

Meanwhile the infantry and the scout tanks will clear up the area.

Third Wave. The object of the third wave is to keep the

battle fluid—that is, to isolate if possible the enemy's anti-tank bases, to cut them off from his mobile forces, to prevent these forces rallying, and to attack his command. It should consist of reconnaissance, destroyer and scout machines, and be followed by the general reserve. Should the enemy be badly shattered, the pursuit should be relentless, no time being allowed him wherein to establish anti-tank bases, or tank-proof areas, from which to recommence operations.

[Notably the case in the German pursuit of the French in June, 1940, and also in Montgomery's pursuit of General Rommel in November, 1942.]

(2) **Attack on a zone of works.** I will describe the nature and defenses of an anti-tank zone more fully in lecture XII. It may be compared to a net opened out in which each knot is an anti-tank work and the meshes in between are lines of fire.

[This has been called "web defense" by F. O. Miksche in his book "Blitzkrieg" ("Attack" in the U. S. A.)]

If these works are tank-proof, as they should be—that is to say, if they cannot be overrun by tanks, or seriously damaged by tank weapons—some other means besides tanks must be thought out to subdue them.

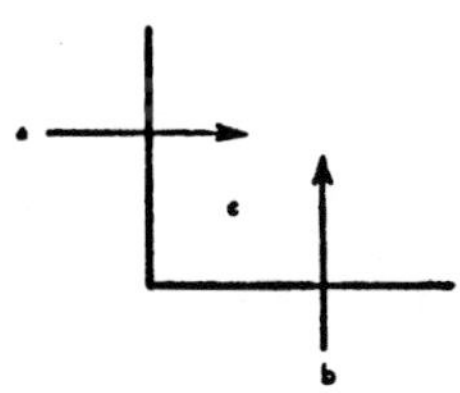

Diagram 16. The Dual Attack.

In all attacks of penetration time is the decisive factor, because in such attacks, unless the reduction of defensive works is more rapid than their construction, penetration becomes impossible. Reduction should, therefore, be combined with forcing an evacuation, and this can best be effected by a dual attack on a salient, the idea being to pinch the salient out. Thus, in

Diagram 16, a single attack delivered from **a** will scarcely threaten the communications of area **c**; but should another attack simultaneously be launched from **b**, then the area in rear of **c** will be placed in jeopardy; and should the attacks from **a** and **b** be pushed far enough, then area **c** will have to be evacuated.

[For attack on salients see my "Machine Warfare," Part ii, Chapter X, and my "On Future Warfare," Chapter iii. Several of the salients in Russia were sufficiently large to constitute major theatres of war. In such cases the above seldom applies.]

As regards reducing anti-tank works, two methods suggest themselves: either they will have to be demolished by artillery or rendered untenable by gas attack. Of these two the second will, in all probability, prove the more speedy, especially if vesicant chemicals are used.

Should it be found impossible to break through an anti-tank zone, then it is certain that offensive operations will be translated to the air, every effort being made to demoralize the enemy by attacking his cities, industrial centers and civil population. Whereas mobile warfare means the attack on armed forces, static warfare can lead to but one end—attack on the civil will.

[This has certainly been proved true. So far as we were concerned, directly the Germans occupied France, the English Channel induced static warfare. The immediate result was the air bombardment of England. The same is likely to happen to Italy once the Anglo-American forces are firmly established in Northern Africa.]

THE FINAL PHASES

75. General Considerations

Should a penetration on a wide front prove successful, the defender will certainly find himself in a critical position, for not only will he have to hold, or beat back, his opponent's mobile forces, but simultaneously he will have to withdraw the garrisons of his anti-tank zone, or, if hard pushed, abandon them altogether.

[This was the difficulty the French found themselves in once the Germans had penetrated their front at Sedan, for they were unable to withdraw the garrisons of the Maginot Line.]

Should the situation warrant it, his best course will be to take up flanking positions at right angles to the enemy's line of advance, and so compel him to turn left or right instead of moving straight ahead. If he can do this, then he may be able to gain sufficient time to throw the flanks of the penetrated defenses back—that is, refuse them—and simultaneously construct a zone of works across, or slightly beyond, the mouth of the gap.

THE PURSUIT

76. General Considerations

In mechanized warfare the pursuit will depend upon whether it follows a field or a siege battle. In the first case the aim of a pursuing force will be to head off the retiring enemy, and so prevent him from establishing an anti-tank base wherein to refit and refuel.

[This is what General Wavell did in January, 1941. He headed off Marshal Graziani's remaining tank forces by a flank pursuit, smashing them in the battle of Beda Fumm. (See "Machine Warfare," pp. 157-158.)]

In the pursuit, as understood today, the first object of a retiring army is to withdraw its transport, and to do so the enemy's advance must be slowed down. In the pursuit of tomorrow the same difficulty will arise, for the main problem will be to establish an anti-tank base well in rear of the withdrawing army, and to do so will generally mean that the enemy must be delayed. We arrive, therefore, at this conclusion: that the pursuing force should be organized in two wings, one to hold on to the enemy's mobile rear guard, and the other to outflank it and cut in between it and its anti-tank base instead of attempting to attack that base.

[General Wavell's pursuit of Graziani in January, 1941, is a perfect example of this. Whereas one part of his pursuing force kept to heel of the Italian retreat along the coastal road, another part cut across the Mekili desert track and interposed itself between the head of the retreat and its base. It was also an identical pursuit by General Grant in April, 1865, which brought the Confederate Army under Robert E. Lee to ruin at Appomattox Court House.]

Here, again, the object of the mobile troops is to attack their like and avoid attacks on anti-tank troops and defenses.

If the pursuit follows a successful attack of penetration, then the best course would appear to be, not to attempt to cut off the enemy's retiring mobile forces, but instead to hold them, and, under cover of this operation, switch forces right and left of the area penetrated, and by severing their communications compel the surrender of the works in the anti-tank zone. My reason for suggesting this is that, if the bulk of the equipment of this zone can be captured, it may be impossible for the enemy to replace it.

[The loss of Tobruk in June, 1942, as well as of the entire rear area of General Ritchie's Eighth Army was disastrous on account of loss of equipment rather than of ground. Had the distance between Tobruk and Alexandria been less, it is highly probably that, for lack of equipment, General Auchinleck would have lost Egypt. Space saved him as time and again it has the Russians; for space is one of the deadliest enemies of supply.]

If so, then he will be deprived of the power to re-establish siege warfare, and will, consequently, have to rely upon field battles to win his cause—battles with no firm base to operate from.

LECTURE XI

CHAPTER VIII

The Defense

77. General Principles

In the history of war we are often told that the passive defense inevitably leads to ruin; yet I know of no writer who has laid it down that the unlimited offense—that is, the French **offensive a outrance**—may in nine cases out of ten be bracketed with it. The reason for this is that the art of fighting depends upon the closest combination of the offensive and the defensive, so closely as does the structure of a building depend upon bricks and mortar. Because the defensive is the less spectacular form of fighting, it is apt, and particularly so during peace time, to be overlooked; yet it is as necessary to correct offensive action as is the bow to the arrow. It is for this reason that time and again I have accentuated the necessity of a strong tactical base for all mobile offensive actions, and if I have seemed to exaggerate its importance it is an exaggeration in the right direction.

As the art of correct fighting depends upon the closest combination between defensive and offensive action, then, conversely, successful battle depends upon how far this combination can be maintained. It may almost be said that victory hinges upon maintaining it, and defeat upon being unable to do so. Therefore, generally speaking, the object of the fight is to separate the enemy's offensive action from its defensive base, in other words—to deprive it of its foundations.

What may be called the defensive order was pre-eminently the tactical order of classical and medieval warfare. Alexander's cavalry were based on the hoplites (heavy infantry) and the Roman hastati (infantry) were based on the triarii (veteran reserve). So also in the Middle Ages were the knights based on their castles or wagon laagers, until Ziska's Wagenburg (wagon-fortress) became an all but unattackable harbor of refuge for his mounted men. So also in mechanized warfare, I believe that a return to Ziska will prove a most profitable study, for the secret in mobile warfare is to remember that, whenever we contemplate offensive action, we must first think of the defensive factor, and that, whenever we contemplate defensive action, we must first think of the offensive factor. This holds good for all operations—marches, halts, field battles, sieges, convoy work, withdrawals and pursuits. In brief, that there must always be a shield as well as a sword, and in mechanized warfare this shield is represented by what I have called the anti-tank base in field operations, and the anti-tank zone in siege.

[So far in the present war all belligerents have to their detriment neglected this.]

It is true that a defensive action is sometimes assumed in order to **avoid** battle, but not to avoid it absolutely, only temporarily or locally. To avoid it in order to resume the fight under more propitious circumstances; or to avoid it in one area so as to battle the more fiercely in another. Such avoidance stabilizes a battle—that is to say, it supplies it with a future, if not an immediate foundation. Hence, let us remember, ever and always, that the defensive is the base of the offensive; hence, let us never forget that a judicious defensive is the foundation of victory.

78. Choice of a Defensive Position

What is the object? This is the question we must always ask ourselves when confronted by a choice. In this problem—namely, the choice of a defensive position—the object is one of three:

(1) To supply a base for offensive action.
(2) To stop the enemy's movement absolutely.
(3) To halt the enemy for the time being.

I have already in previous lectures dealt **ad nauseam** with the first of these objects, so here I will examine only the second and third.

In the second, the position, or area, selected should be such that it can only be attacked frontally; in the third that, when the enemy attacks, his rear, or flanks, will be laid open to attack. Once again flank and rear are the points which should guide our decisions and direct our thoughts when we think out our defensive plan. In both cases, observation of the enemy is the most important "reinforcement"; for as choice is largely restricted by the nature of the locality in which we are fighting, and as few localities will offer us an ideal defensive area, the more we learn about the enemy's movements, and through his movements of his intentions, the more fully shall we be able to defend ourselves.

Now, in present-day warfare, artillery is the defensive base of infantry action; consequently a defensive position should possess good artillery O.P.'s, because laying is indirect. But in mechanized warfare ground observation is nothing like so important as air observation; for the base of the mobile troops—the tanks—is not artillery but the anti-tank troops. These troops will seldom resort to indirect laying, for to hit their targets they must see them. Though ground observation should never be neglected, it is not so important as discovering the direction the enemy is moving in. Once this is known, the defenses can, if necessary, be reshuffled, for with mobile defenses reshuffling is not an impossible task.

[No army has yet tumbled to the fact that, so far as possible, defenses should be mobile. This is a mobile age and mobility is its key.]

Therefore air power is of vital importance to the defender, and only second to it is the maintenance of contact with the enemy by armored cars and motor-guerilla forces. Therefore also, it is advantageous should a defensive area include a good landing ground. In fact a landing ground is more important than a number of excellent artillery O.P.'s.

When a defensive area is chosen to stop the enemy absolutely, a salient, or salients, should be avoided; but if only temporarily, then a salient, so I think, is an advantage, especially if a defensive chord is constructed across its base. This is expiained in Diagram 17. If **a b c** is the salient, the enemy, **d,** can attack either **a-b** or **c-b** or **b.** Should he attack **a-b** his left flank can be struck at from **b-c** as shown; if **c-b,** then from **a-b,** and if **b** then from **a-b** and **c-b.**

In selecting a defensive area, obviously the ground must be examined in relationship to the object. If it is to stop the enemy, then its anti-tank possibilities demand our first consideration; if it is to halt him temporarily, then its facilities for eventual tank

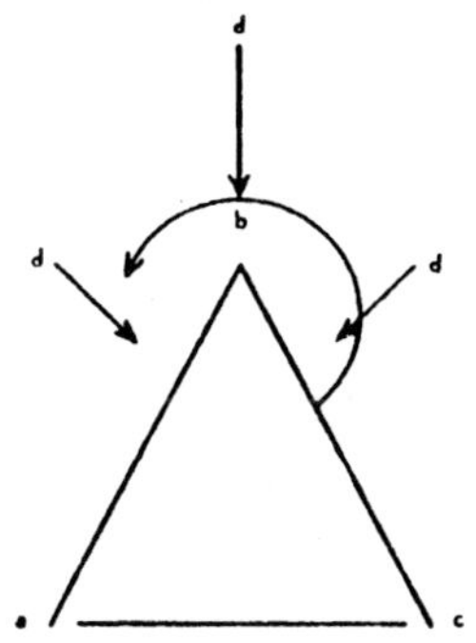

Diagram 17. Defense of A Salient.

movement are of first importance. In the first case a good field of fire should be sought, not only because the enemy's machines can then be engaged at long range, but so that anti-tank works may mutually support each other. If possible, anti-tank obstacles should flank and run immediately in front of the area, such as a stream or sunken road.

[Minefields should be added to this list.]

If such obstacles cannot be seen by the enemy, then they will come as a surprise to him. As, generally speaking, this combination of obstacles will not be found, I consider it likely that valleys and defiles will play an important part in future defensive warfare, for when the defender can rest his flanks on natural obstacles, all he will then have to consider are the artificial defenses on his front. This type of defense was predominant in Greek Classical warfare.

[This is largely true of the British position at Dunkirk in May, 1940, though there the flanking obstacles were two inundated rivers—the Aa and the Yser. Their flanks secure all the British had to think about was their front which was secured by numerous dykes uncrossable by German tanks until bridged. (See also Last Annotation Page 127.)]

79. Preparations for Defensive Action

The foundations of all preparation are laid by discovering the enemy, or, if he is already discovered, in maintaining air and ground contact with him.

Defensive localities fall into two groups—the natural and the artificial. In the first we obtain hills, rivers, swamps, woods and villages; and in the second, inundations, mine fields, anti-tank trenches and anti-tank works.

Mine fields, as I have already stated in a previous lecture, are to tanks what wire entanglements are to infantry—the one stops the mobility of the machine, the other of the man.

[Such as the Gazala-Bir Hakeim and El Alamein minefields.]

A mine field should consist of rows of mines laid diagonally, its object being to bar movement or to deflect movement. Generally speaking, mine fields should be protected by fire—that is to say, they should be used in conjunction with anti-tank weapons. To prevent men and animals walking over them, they can be fenced in or covered with entanglements. It is true that such precautions may disclose their position; nevertheless, this disadvantage is in part compensated for by the fact that many dummy mine fields can be laid out in order to mislead the enemy.

Anti-tank entrenchments may take three forms—namely, banks which cannot be climbed, trenches which cannot be crossed and narrow trenches to entrap the tank tracks. The first are suitable obstacles where steep slopes exist, such as the sides of a sunken road, a railway embankment and the banks of a river. They are easily made by cutting the slope down until it presents an all but vertical surface six to seven feet high. The second are a definite protection, but as they will have to be at least twenty-five feet wide and ten deep, they take a long time to dig, and can scarcely be considered practical defenses except in the case of permanent fortifications. For the protection of field works, the third are the easiest to construct, and consist in rows of slits, or graves, ten feet long, two feet wide and three to four feet deep, set sufficiently close together so that should a tank attempt to cross them, one or both of its tracks will sink into them, the machine becoming bellied, or ditched.

Finally, we come to anti-tank works of earth, concrete or steel. I have already touched upon these, but here I will enter into more detail. They may be divided under two general headings —the fixed and the transportable. The first will take the form of the old Martello tower protected by a moat, a ring of anti-tank trenches, or a mine field, and in some cases all three combined. These towers will generally be much lower than their famous ancestors, and if surrounded by a moat will only just peep over it.

[Had the three airfields in Crete been defended by such works, it seems doubtful whether the German air force would ever have been able to occupy them.]

They may be built of earth revetted with hurdles and gabions, or constructed of concrete, and will, in all probability, be armed with one high velocity anti-tank gun mounted in a revolving cupola, and two anti-tank machine guns; these can be used for anti-aircraft defenses as well. To man such a stronghold would not require more than a commander and nine men—about the same garrison as the circular blockhouse of the South African War.

As regards the transportable defenses, these are likely to take two forms. The first, a bullet-proof steel cupola carried in sections, a somewhat heavy arrangement, and the second semi-circular pieces of ordinary corrugated iron which can be set together in two circles, like the walls of the South African block-house, the space in between being filled with gravel or earth. These anti-tank pill-boxes—to distinguish them from the cupolas—will not only be light to carry but cheap to manufacture. Being roofless, the anti-tank weapon used in them should be equipped with a slanting semi-circular shield.

80. Organization of Defense

The organization of the defense will obviously depend on the character of the country, and the natural obstacles to be found in it. Vital approaches, rather than vital points, will be the dominating factors; therefore before he decides on his defenses, the defender should carefully look to his flanks and rear.

The general scheme of defense in every case should comprise a system of anti-tank works and a mobile counter-attacking force generally kept well in rear and towards the most exposed flank, so that, should the enemy attempt to turn the works, he can be attacked in flank or rear.

As regards natural obstacles, the most important are woods, rivers and villages.

Woods should be defended by anti-tank machine guns, supported by artillery. Though if of a dense nature they constitute definite obstacles to tank attack, they frequently form admirable covered approaches and lying-up places. Should tanks be allotted to support their garrisons, they should be so placed that, when the enemy is held by anti-tank fire, they can attack him in flank.

To all but amphibian machines rivers also form definite obstacles; therefore, their importance is likely to grow in defensive warfare.

[This would certainly be true in Russia.]

They should mainly be held by anti-tank machine guns supported by artillery, and when tanks are allotted to assist in their defense, they should be kept concentrated in such positions as will enable them to counter-attack the enemy directly he has effected a crossing.

[In 1940 the French failed to do this along the Aisne and Somme rivers, and the result was that the Germans had a free run into Paris.]

They should not be split up into small packets to watch bridges, fords, etc. When such are to be defended anti-tank weapons should be used.

Generally speaking, villages are unsuitable localities for tanks to fight in, because movement is restricted to the streets and roads.

[See First Annotation Page 131.]

When tanks are allotted to assist in the defense of a village, they should, if the village is large, be held in reserve in a central position; but should it be small, then in rear or on the flanks, hidden away from observation, and so placed that they can attack, or counterattack the enemy in flank when held up by, or moving away from, the fire of the anti-tank garrison.

81. Occupation of a Defensive Position

As I have just ponted out, in the defense two forces are always required—namely, the holders and the hitters; and in a mechanized army, as I have shown throughout these lectures, this division automatically divides the troops into anti-tank and tank forces.

Each defensive area is garrisoned by these two, and, therefore, should be sub-divided into mobile and static tactical sectors; the second canalizing the attacker, and so leading him into areas where he can be speedily counterattacked by tanks. Here is presented to us a somewhat novel conception of the defense—namely, that a counterattack is not necessarily an attack on an enemy who has broken through a defensive system, but instead has been shepherded by the defenses into areas where he can be attacked with advantage. I will deal with this spider and fly defense more fully in my next lecture.

As regards the static defensive areas, as I have pointed out in a former lecture, they should consist of belts, or groups, of mutually supporting works, and the inner flanks, between the units garrisoning them, should, I think, run diagonally and not perpendicularly to the probable frontage of the attack, so that each unit overlaps the works of the unit on its right or its left. This will automatically strengthen them.

If the defensive area is surrounded by open country, then, obviously, the defenses will have to be of an all-round nature; in which case the mobile forces should be kept well concentrated and ready at a moment's notice to strike at the enemy should he attempt to cut off the area from its base of supply.

As regards the general distribution of weapon-power, whether in works or in the open, it should be governed by the principle that the anti-tank machine guns should be so sited as to compel an attack to move broad-side on to the artillery defenses, and that the fire of these defenses should be so directed that they will drive the enemy's machines towards the counterattack tanks.

82. Counterattack and Counteroffensive in Defense

I have already mentioned the counterattack, and have pointed out that defenses and weapons should be so sited that they will facilitate it being delivered in a **selected** area— that is, not only an area which will be advantageous to the counterattacking troops, but which will enable the counterattack to come as a surprise, and the troops taking part in it to be protected by the anti-tank defenses. As the anti-tank machine guns should be so sited that their fire will drive the attacking machines onto the defender's artillery, and as the guns should be so sited that they will drive them into the counterattack areas, so should the counterattacking force be so placed that, when it attacks, it will drive the enemy's machines onto a well concealed mine field. Here we have presented to us a picture of the closest co-operation between the various defensive and offensive arms, and this picture must be present in a commander's mind at the time when he decides upon his defensive arrangements.

To turn now to the counteroffensive. Should an anti-tank zone have been constructed during peace time, then everything is to be gained by inducing an enemy to attempt to penetrate it.

[It will be noticed that the reverse was the case with the Maginot Line, the underlying idea of which was that of a wall—it must not be pierced.]

Similarly in field warfare, as a mechanized army will consist of two forces—tank and anti-tank—whenever it is possible for the latter to take up a position which the enemy will have to attack in order to carry out his plan, then it will generally be to the advantage of his opponent to let him attack, and directly his attack begins to succeed, or fail, to launch a counteroffensive in full force against him.

[For instance, had General Rommel at the Battle of El Alamein established three or four anti-tank fortresses in rear of and covering the flanks of the sector General Montgomery was attacking, once the British Xth Corps was through the minefield, it would have been faced by them as well as Rommel's three armoured divisions, which could have used them as pivots of maneuvre, In May-June, 1942, the clumsy "Knightsbridge Box" was sufficient to turn Rommel back after his first penetration of the Gazala-Bir Hakeim minefield.]

The whole question of the counteroffensive is one of common sense, and also one which will bulk large in mechanized warfare. If you can persuade your adversary to blunt his sword against your shield, indirectly so much the sharper becomes your own sword. As both sides are extremely mobile, when compared to infantry in the attack, both sides can advance and withdraw at will. And as both possess tank and anti-tank forces, it is obviously to the advantage of either to persuade its enemy to measure his strength against its anti-tank forces—which are specially designed to resist mobile attack—before launching its tank forces against him. In this respect we may talk of an offensive-counteroffensive—that is, an advance to engage the enemy, a withdrawal to persuade him to advance against anti-tank defenses, and then again a forward movement to strike him before he can reorganize. Such was William the Conqueror's plan at the battle of Hastings.

83. Infantry in Defense

The part infantry will play in the defense, other than the motorized infantry which will assist in manning the anti-tank defenses, will depend on the nature of the ground. If unsuited for tank movement, which will generally mean that they will be engaged in forest and mountain fighting, then their tactics should be those of light infantry. Should they be equipped with anti-tank weapons, then, normally, they should be kept in reserve, so that they may freely be moved towards any threatened point.

84. Artillery in Defense

Artillery in the defense will include static guns in the anti-tank defenses and the mobile reserves of artillery tanks. The object of the first I have already dealt with, that of the second is to reinforce any threatened point, and more important still to protect the counterattacking troops by coming into action in such positions that the enemy's power to maneuver will be restricted.

[On several occasions General Rommel would appear to have used his self-propelled 75 mm guns in this way.]

Again, it will be seen here that protection is indirect—that is to say, its object is not to reduce the enemy's offensive strength, but to restrict his power to maneuver. These mobile guns, as I have already explained, will normally fire from a stationary position, in order to attain higher accuracy of fire, while the scout, combat and other machines will fire in movement, and if possible drive the enemy towards the tank artillery.

85. Cavalry in Defense

Cavalry can play no part in the defensive operations of the future. They are not suited either for mechanized warfare or for mountain or forest warfare. Their present duties will be taken over by armored cars and armed motor cars. The first will gain contact with the enemy and keep him under observation; the second, when he engages, will harass his flanks and his rear.

86. Engineers in Defense

Field engineers should be used in the construction of anti-tank works which cannot be built by the anti-tank troops. Their main duties, as today, are likely to consist in bridging, water-supply and demolitions. As specialists they should be kept for special work. The mechanical engineers will carry out their normal duties of repair work in the field.

[Such a Corps, first recommended during the last war, was created in Great Britain in 1942.]

LECTURE XII

CHAPTER VIII (Continued)

The Defense

FURTHER CONSIDERATIONS IN PROTRACTED DEFENSE

87. Special Conditions

In mechanized warfare the protracted defense may at first seem an anachronism, but, as I have already shown in previous lectures, mobile warfare will inevitably lead to static warfare, and possibly to warfare as static as that experienced during 1914-1917; yet with this difference: That whereas then entrenched line held entrenched line, in static mechanized warfare fortified zone will hold fortified zone. There may be more elasticity of movement, but it will be within zones, the territories outside of them being as fully protected from ground invasion as they were during the greater part of the World War.

[It would appear that this is what in 1941 happened in Russia. The fundamental German mistake or misfortune was that Russia was not overwhelmed in a single campaign. To do so now demands new tactics, which in turn demands either a powerful new weapon or a surprisingly novel use of existing weapons.]

It may be considered that such complete protection will be prohibitively costly. It is true that the cost may deter a nation building esuch defenses during peace time; but that it will prove as costly as the protection established during the last war is unlikely, seeing that the defenses then used were continuous, that they required thousands of guns, tens of thousands of machine guns and hundred of thousands of infantry to hold them.

If they are constructed, then we may be certain that they will prove sufficiently costly to prohibit any nation maintaining a gigantic mechanized army as the projectile to be shot from them.

[There can be little doubt that the cost of the Maginot Line was one of the main reasons for neglect of mechanization in the French Army.]

In fact, here is presented to us a somewhat curious picture. The tank, an essentially offensive weapon, will at once react upon the protective idea, forcing it into first consideration. And this protective idea, leading to the construction of elaborate zones of defense, will, through cost, react upon the offensive idea by limiting the number of offensive weapons.

There is also another reason why the protracted defense is likely to bulk large in the future. Once again it springs from the tactical influence of the gasoline engine, not in the form of the tank, but in that of the motor vehicle. As I have shown, motorized guerillas are likely to be used in large numbers, because they can so easily be raised. These troops, as I have pointed out, are in nature essentially defensive, and though when fighting in their own country, they have nothing to fear from the civil inhabitants, this security begins to crumble directly they enter the enemy's country, and the further they enter it the less will they be able to protect the mechanized forces they are intended to cover, and the more will they depend on these forces to protect them, until it is possible that they may become an encumbrance rather than an assistance.

Should such a situation arise, or should they be withdrawn, the mechanized forces will be left to fend for themselves—that is, to find their own protection. Should these forces number thousands of machines, this problem would not be a difficult one to solve; but as so large a number is unlikely, to fend for themselves in face of the enemy's guerillas, and above all to prevent the area of communications from being attacked and possibly closed, will place such a strain on them that they may be forced to halt. To halt is to stand still, and prolonged periods of stand-still fighting must lead to the protracted defense.

The solution to this problem is to provide the guerilla forces with a protected base, so that as they and the mechanized army advance, they may find in their rear, if not a friendly then a peaceful and not a hostile country. In fact, if mobile warfare is to succeed, the fighting forces, motorized and mechanized, must have in rear of them an administrative base, that is also a strategical base, as there can be no strategy without adequate maintenance, supply and evacuation. This brings us to the problem of the army of occupation.

Whereas the tactical bases, of which I have so often made mention, are areas temporarily held for purposes of battle, the strategical base is one permanently held for purposes of victory —the ultimate result of a number of battles. This strategical base is flanked on one side by the common frontier, and on the other by a "wall" of troops which advances under the protection of motorized and mechanized forces and establishes law and order in the territories occupied. Thus, by degrees is the enemy's strategical area reduced, until, so to speak, it is pinched out of tactical existence.

From this it will be seen that there are two supremely important grand tactical problems in mechanized wars. The first, to keep the mechanized and motorized forces moving, so that the enemy's country may be occupied; and the second, to prevent this occupation by compelling the mobile forces to assume a protracted defense. In brief, the whole problem is one of moving and stopping movement, in order to occupy, or prevent occupation. When this is fully realized, there can be no question that anti-tank defense will bulk large in the military problems of all continental general staffs. Further, once these staffs are satisfied that their frontiers are tank-proof, they will realize that they form excellent bases for air attack, which today would appear to be the ultimate problem in mechanized wars—mainly, how to maintain offensive mobility from a protected static base.

88. Organizations for Protracted Defense

There can be no doubt, if history is consulted, that an offensive war—that is, a mobile one depending on field operations—is far less costly and destructive than a defensive one. In the World War it was because the bullet on the defensive was a more powerful weapon than the bullet in the attack, that the war proved almost as disastrous to victor as to vanquished. Had it been otherwise, that war would have ended in a few weeks, and its cost and destruction would have been reduced proportionately.

THE DEFENSE

As long as motorized and mechanized armies can maintain their mobility, wars are likely to be short, though possibly not so short as some people think, for it must not be overlooked that, though mobile operations may be brought to a halt on the ground, it in no way follows they will simultaneously be brought to a halt in the air.

[This is what has happened. The rapid German conquest of the greater part of Europe did not bring air warfare to an end, rather did it stimulate it. As air encirclement of Great Britain was one of Germany's main tactical aims, so it would seem that the encirclement of Western Europe is now one of the main aims of the United States and Great Britain. In other words—blockade by terror.]

If I am right in supposing that protracted defensive warfare will mean fighting in the anti-tank zones, then we may be certain that the object of this fighting will be not so much to penetrate the enemy's zone as to keep him fixed in it, so that a secure air base may be established, from which a highly organized air offensive can be launched against the enemy's industrial centres.

[As I write (spring, 1943), something like this would appear to be developing on the following lines: (1) A German defensive both in Russia and the occupied countries, under cover of which attack on British and American shipping will be intensified; (2) an Allied defensive on the seas, under cover of which from air bases established in North Africa, air attacks, will be launched on the Southern half of Western Europe, whilst similar attacks are launched from Great Britain on the northern half. The two great three-dimensional weapons—the submarine and the aeroplane would appear to be in clinch.]

From what I have already said in former lectures, it is of course possible for an air offensive to take place at any time, but with this difference: at the beginning of a war it is obvious that no nation is going to allow its territories to be overrun by swarms of motorized guerillas and mechanized forces. Either during peace time it will have constructed an anti-tank zone which will stop them, or it will fight them in the field. In the first case a protracted defense may take place immediately; in the second an attempt will be made to defeat the enemy's land forces in the field, and a protracted defense will be established only when this attempt fails. Though in the first eventuality the probabilities are that offensive action will at once be sought through air mobility, in the second it will obviously be sought through the mobility of the land forces, air forces cooperating with them and seldom acting independently unless the air power of one side vastly outbalances that of the other.

[This has proved correct: protracted defense has led to air attack, and mobile land offense has demanded the closest possible cooperation between land and air forces.]

It is generally overlooked that though in the air aircraft are the most mobile of weapons, on the ground they are the least mobile; in fact they are not weapons at all. Further still, their whole ground organization—hangars, workshops, landing grounds, etc.—is highly immobile, so much so that unless this organization can be guaranteed immunity from land and air attack, its utility is vastly reduced. What a highly organized and sustained air offensive requires are the conditions which prevailed during the greater part of the World War—namely, an entrenched front without open flanks, and immunity of attack behind this front. Given this security, then and then only can aircraft develop to the full their offensive power; not given it, then, because their bases of action will either be insecure, or distant from the enemy's industrial centres, much of this power will be lost.

[On account of the ever-increasing range of aircraft, this is becoming less and less true.]

We thus see that in independent air operations—attacks on the civil population—the first thing to do is to establish a protracted defense.

I will now turn to the anti-tank zones, which I have already discussed in some detail; nevertheless, there are several points which require elaboration. The zone itself will consist of all such territory which if lost will entail surrender. This zone may be as much as 200 miles deep. Obviously, therefore, it is out of the question to cover it with a network of anti-tank defenses. I consider, therefore, that the main zone will be divided into two sub-zones, which I will call zones **A** and **B.** **A** will consist of a belt of works bordering the frontier, and **B** of the remainder of the main zone. **A** may be compared to coastal fortresses and **B** to the old medieval castles which blocked inland approaches; for in it all strategical centres will be defended, such as large railway junctions, industrial districts, aerodromes and the capital. Should zone **A** be penetrated, then, though the invader will not be confronted by belts of works, his movements will be restricted by these anti-tank and anti-air castles.

[It will be seen that this system of defense is very different in idea from the Maginot Line and similar Great Walls of China.]

Not only will they form harbours of refuge for his opponent, but in most cases he will be compelled to reduce them before moving forward, for to pass them by will lay his area of communications open to constant attack.

In brief, the protracted defense in mechanized warfare is brought about by entangling the enemy's forces in a tactical net of fine mesh from which, should he break through it, he will find himself caught in a strategical net of wider mesh. Should he refuse to enter either, a clinch will take place, and in both cases offensive mobility will be sought in the air, the target being the civil foundation of the contending armies.

89. Counterattacks in Protracted Defense

The counterattack in the protracted defense in no way differs from that in the field defense, except that the odds more greatly favour the counterattacking forces, because the defensive base they operate from will be stronger. As we have just seen, the main counteroffensive will be relegated to the air force, and will consist in first gaining air supremacy, and secondly in selecting one or more vital targets, the enemy's capital, or one of his industrial centres, and by keeping it under prolonged air bombardment, force its evacuation instead of its surrender, and then switch onto another target.

90. Infantry in Protracted Defense

The bulk of the infantry forces in the protracted defense will be the army of occupation. Unlike present-day infantry, their duties will be static instead of mobile, and their main weapons of defense will not be the rifle and the machine gun, but the anti-tank machine gun and land mine. Should an anti-tank zone exist, on the outbreak of war they will occupy it; should it not, then they will have to construct it. In the area selected for its construction their first duty is to stop the advance of the enemy's guerilla swarms, and should the enemy's mechanized force be driven back, their next is to push forward the anti-tank zone, and thus step by step wall off sections of the enemy's country in order to establish a state of security in rear of the advancing army.

[It would appear that this is the present intention of the Germans in Russia.]

91. Artillery in Protracted Defense

Omitting the guns of the mechanized army, the artillery of the protracted defense will be provided by the army of occupation. There are likely to be four main classes of guns and howitzers—namely, siege, field, anti-tank and anti-aircraft, as well as special cannon, such as Big Bertha of World War fame.

92. Cavalry in Protracted Defense

I can see no place for cavalry in the protracted defense other than that of mounted policemen and traffic control posts. In all probability motor cyclists and field police carried in light cars will prove more economical and effective.

93. Engineers in Protracted Defense

The main force of engineers will, obviously, be with the army of occupation. It is likely to consist of two categories of troops—mechanical engineers and field engineers. The first will carry out the repair work of all machines of the field army and the army of occupation, and the second will be responsible for the surveying, planning and laying out of the various fortifications in the anti-tank zones.

94. Information in Protracted Defense

Information in the protracted defense is as important as in the field defense, but as motorized forces will seldom be able to operate forward of the anti-tank zone, aircraft will have to be relied upon for all reconnaissance. Works will be linked up by cable and wireless to the section command posts, and these to general headquarters. In fact the whole of the anti-tank zone will, for purposes of intercommunication, be organized like a fortress.

95. Reliefs in Protracted Defense

It is unlikely that the system of reliefs as generally adopted during the World War will be resorted to in the type of warfare with which I am now dealing. Then a wall of men manned a flankless belt of entrenchments, and their main object was tactically a rigid one—namely, to prevent this belt being penetrated. The strain placed upon the trench garrisons was great and to maintain their morale and physical efficiency reliefs were frequent and by complete units. In future this human wall will be broken up into a large number of small anti-tank garrisons; therefore the anti-tank belt will, so to say, be all flanks, but of great depth.

[So far as can be gathered, this is the German system on the greater part of the Russian front. There is no continuous line, therefore break-throughs are more frequent than in 1914-1918.]

Though the object of these garrisons will be to prevent the belt being penetrated, on account of its depth there will be far more room to maneuvre in it. When the enemy attacks a post the object of the garrison will not generally be to die in it, but to get out of it. Of course, should the post be one of vital importance, it must be held staunchly, but if not, after the enemy has deployed his force, the garrison should retire and occupy another post at the further side of zone **A.** Thus the defense will be extremely flexible, each bulge made by the enemy resulting in an automatic thickening of the defenses **in rear.** Reliefs will, therefore, depend on the enemy's action, and will become more or less automatic when he attacks, and cease to take place when he does not, as is the case with the garrison of a fortress.

DISENGAGEMENT

96. General Principles

Disengagement in the field will depend on how far liberty of movement has been retained, which itself will depend on the

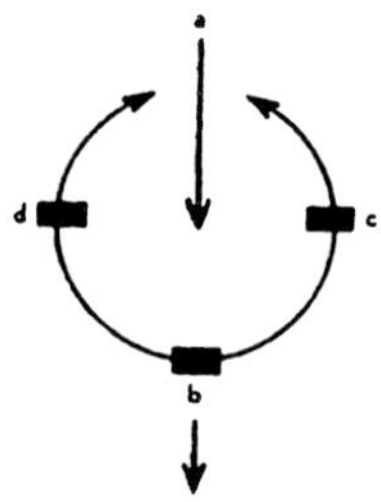

Diagram 18. Disengagement

strength of the reserves. If order has been maintained—that is to say, if forces have not become scattered—and if reserves are strong, a disengagement should not generally be a difficult operation. The normal action will be to advance the reserves on one or both flanks of the force to be disengaged and, by threatening to attack the enemy's flanks, compel him to slow down his advance. Thus, in Diagram 18, if **a** be the enemy and **b** the force to be disengaged, the movement of the reserves is shown as **c** and **d.** Once the withdrawal has begun, **b** will fall back towards the anti-tank base whilst **c** and **d** protect **b's** flanks. Once arrived at that base, **b** will refit, whilst half the base packs up and moves to a position in rear. Then the remaining half will retire protected by **c** and **d,** and so on, **b** eventually relieving **c** and **d** of the rear guard work.

In the case of a withdrawal from a defensive position, or a section of it, the problem will be more difficult, because the defensive garrison will not be highly mobile. Again the principle is the same—the mobile units covering the retirement. When a post has to be evacuated, the best time to do so is obviously during the night.

LECTURE XIII

CHAPTER IX

Night Operations

97. General Principles

There can be little doubt that the difficulties inherent in night operations will apply to mechanized armies as fully as to those of the present day, and that, with the exception of aircraft and motorized guerillas, they will be no more frequent than they have been in the past.

[This has been proved correct, except in North Africa where open and unrestricted space facilitates night movements.]

As regards the motorized guerilla, if possible he should be trained for night work, because darkness will be his surest protection, and as he will be able to move great distances at a stretch, night will enable him to change his position from flank to flank, flank to rear, etc., by many miles. Unless he knows the country intimately, his chief asset will be not so much map reading as map memorizing.

With infantry, night movements are frequently undertaken to avoid the enemy bullets, but except in the case of an attack on anti-tank defenses, this reason will seldom be the main one for armoured troops. Night movement by them will generally be carried out, not so much to shield themselves but their intention. By shifting their position at night they will frequently be able so completely to upset the enemy's information that he will show extreme caution during the early hours of the next day.

[In North Africa, so I understand, this form of protection has become routine. As night falls armoured troops go into close laager, and before sunrise either shift their position or go into open laager. In other words, they contract and expand.]

The result of being able to move considerable distances without being seen is likely to introduce what may be called "the tactics of hide and seek," and, therefore, will bring the feint, the ruse and the ambush back into their own. Again, this will considerably slow down field operations, otherwise a mechanized force may at any moment find itself in the position Flaminius found himself in at Lake Trasimene.

Personally, from what I learnt during the World War, I do not think that these night movements will be as difficult as many soldiers believe; especially if tanks are equipped with a reliable compass, and the training of their crews has been reduced to a routine.

[In North Africa this has proved correct.]

As regards night attacks I see no reason to suppose that they will become any more frequent in the future than in the past, and I consider that they will seldom be resorted to in order to break through an enemy. This will not, however, prohibit attacks being made for moral instead of physical effect, for at night time, though the physical effect of tanks is frequently negligible, their moral effect is considerable. By using them for this purpose an enemy is almost certain to be kept in a state of anxiety, and if the initiative is with his opponent he will be kept in constant suspense.

We may, therefore, conclude that, whilst the physical attack during hours of darkness will be resorted to only when no other means is likely, to succeed, the moral attack will become a fairly common operation in future warfare, its object being to exhaust the enemy's nerves by depriving him of rest and causing him constant anxiety.

[I am of opinion that this is more especially true of air attack during hours of darkness: moral effect is greater than physical.]

NIGHT MOVEMENTS CARRIED OUT BEHIND PROTECTIVE TROOPS

98. General Considerations

As I have explained, in mechanized warfare protective troops may be grouped into two categories—namely, the all-round defense provided by the guerilla swarm, and those of the anti-tank zone. In the case of the latter, night movements will approximate more closely to peace time conditions; but in both the main difficulty will be driving without head-lights, and the main danger must be expected from the air.

99. Conduct of Night Marches

Night marches may be grouped under two main headings: (1) Road marches, and (2) cross-country marches. As regards the first the main danger lies in obstructions on the road—land mines, barricades and enemy anti-tank weapons; and as regards the second, difficulty in finding the way. When such movements are to be carried out, their success will mainly depend on careful reconnaissance. Should the country be broken up by hedgerows, tracks, woods and isolated farms, air photographs will prove of the greatest value, because they enable us to plot a zig-zag course from one feature to another, and the distance between features having been estimated, all the leading tanks need do is to move forward on a compass bearing, changing it as each feature is reached, the remaining machines following their leader. On certain occasions it may be found useful to march tanks on coloured lights, or at a number of degrees to their right or left, these lights being dropped by airplanes or fired by guns. Thus, should the enemy's front extend from **a** to **b** and his rear be located at **c,** by firing red lights on **a,** green on **b** and orange on **c,** great assistance will be given to an attacker, not because he can march directly on these lights, but because they will define for him the position of the enemy. Such means, coupled with air photographs and possibly also wireless direction, should go a long way towards reducing the difficulties of night movements.

A point which should not be overlooked is that of noise. Noises at night are normally most deceptive, they are difficult to follow, for the slightest current of air will alter their direction.

Formations should generally be in line ahead, and seldom larger than a company of machines in each line, or column, flanking contact between lines being maintained by the use of dimmed port and starboard lights. The use of head-lights will normally be forbidden, but tail lights will be used.

NIGHT MOVEMENTS WHEN THE FORCE IS RESPONSIBLE FOR ITS OWN PROTECTION

100. Night Marches

Generally speaking, in all field operations a mechanized force will be responsible for its own protection, which I have frequently stated must be of an all-round kind. The main danger during night, as during daylight, is an attack in flank. A night march may in many ways be compared to a bush warfare operation, for in both lack of visibility and, consequently, liability to be surprised are the main factors. In order to prevent delay in movement, the area occupied by the advancing force will have to be a contracted one, the main body should move in column, or in a series of columns, with small flanking parties on each flank of its advanced guard, rear guard and centre. At a distance from these, and if possible on roads flanking the main line of advance, should move the two flank guards and the patrols.

101. Night Advances

A night advance demands a tactical instead of a logistical formation, as its object is to attack an enemy either during hours of darkness or at dawn. During the World War tank night advances were frequently undertaken, and the most notable was that which preceded the battle of Cambrai. In this particular operation tanks moved from their lying up places to their positions of deployment by following tapes laid on the ground. But such a method of guidance will be the exception in future warfare, as far greater distances will have to be traversed. In these operations the two main difficulties will be to maintain formation when deployed, and, if the advance is a secret one, to prevent the noise of the tanks attracting the enemy's attention. To overcome the first, small compact columns will have to be employed, the scout tanks leading and the rest keeping closed up in rear. To overcome the second by direct means, that is by preventing all noise, is not really practical. At the battle of Cambrai it was accomplished by advancing at less than half a mile an hour, but then it must be remembered that few of the machines had to move in all more than two miles. A better system would appear to be to move as fast as the maintenance of formations will allow, and simultaneously carry out a large number of feints so that the noise of machines over a wide and deep area may confuse the enemy and prevent him discovering the point selected for attack.

The whole question of the direction of noise depends on the nature of the ground, air currents, temperature and atmosphere. It is a problem little understood, and one which might well be experimented in during peace.

102. Night Withdrawals

There will be two main kinds of withdrawals in mechanized warfare—namely, withdrawals to an anti-tank base, and withdrawals from such a base. The first will normally be carried out during daylight, and will constitute either a rapid movement backwards, or a slower leap-frogging back by echelons. The second will best be carried out during hours of darkness by withdrawing half the anti-tank troops well to the rear, and then the remaining half under cover of the mobile forces. As in mechanized warfare withdrawals are likely to be as frequent as advances, two considerations arise: The first is that the anti-tank equipment must be such that it is easily transportable in cross-country machines. The second, that reconnaissances for withdrawals, either to the rear or to a flank, must unceasingly be carried out, so that not a moment is wasted when a change of position is ordered.

NIGHT ATTACKS

103. General Principles

In mechanized warfare, night attacks are unlikely, unless: (1) The enemy's position is virtually unattackable by day; (2) the enemy has been maneuvred into an area from which escape is only possible during hours of darkness.

[This has been proved true. None of importance has been reported, unless the Battle of El Alamein is an instance, because British mine-clearing was mainly carried out at night time.]

In the first case the attack will demand considerable preparation; in the second, generally, a rapid maneuvre, which will lead to the enemy being struck in force as he is attempting to escape.

When a night atack is decided upon, the plan should be as simple as possible, the objectives selected being easy to find and sufficiently far-spaced to prevent the attacking columns converging upon each other in the darkness. As during night time it is always difficult to distinguish friend from foe, the attacking machines should be marked by a distinct and easily recognizable sign.

I will repeat it again, the main danger in these operations is loss of direction and not the enemy's fire power. A hit in the night is nearly always a fluke; but a wrong turning is an ever-present danger. A question which is certainly worthy of detailed trial is the use of search-lights in these operations. At sea search-lights have frequently been used to blind an enemy, and I see no reason why they should not similarly be used in land battles. One thing is almost certain, that they will be used in the anti-tank defenses, as they are to-day in the anti-aircraft defenses; therefore the probability is that the mobile forces will also be equipped with them, which in itself will raise a host of new night warfare problems.

104. Orders for Night Attacks

Orders for any form of night attack should be as simple as possible. Direct movements should be ordered, and the timing of feint movements carefully worked out so that the whole area of attack may be kept under constant alarm until the real blow is struck. Every commander must fully realize the part he has to play and what to do if his part fails.

NIGHT MOVEMENTS BY MECHANICAL VEHICLES

105. General Principles

The chief danger to be expected in the movements of convoys by night is from the enemy's guerillas, as it will be the exception and certainly not the rule for the enemy to move tanks during hours of darkness round his opponent in order to attack a problematical convoy. If no guerillas are about, night is obviously a good time to move in, for even should enemy airplanes attack a convoy their aim will not be so accurate as during daylight. Such movements are, however, always slow, as head-lights can seldom be used.

Should, however, the enemy's guerilla forces be known to be in the neighbourhood, and this will generally be the case once his territories are entered, all-round protection must be maintained. This can best be effected by using two bodies of troops. Thus, in Diagram 19, if **a** is the convoy, then these two bodies are represented by the circles **b** and **c.** As the convoy moves off, **b** pickets and patrols the circumference of the area **d,** and **c** moves forward and does the same for area **e.** As the convoy enters area **e, b's** pickets and patrols move forward to area **f,**

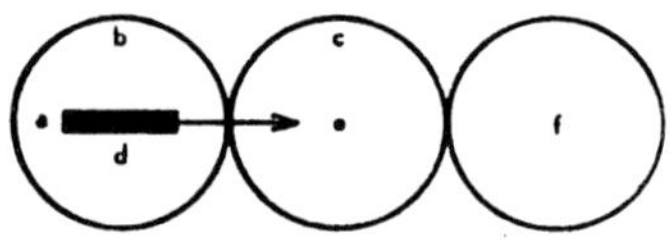

Diagram 19. Protection of A Convoy.

and so on. A road convoy of any size is not likely to move at a greater speed than from seven to eight miles the hour; consequently the problem of how long the diameter of the circles **d, e** and **f** should be will depend on the road facilities on the flanks offered to **b** and **c.** Should a seven mile radius be decided on, then these facilities should be sufficiently good to enable **b** to move from area **d** to area **f** in one hour.

When cross-country convoys are used, this system of protection may be inapplicable, in which case the escorts will have to consist of scout tanks.

LECTURE XIV

CHAPTER X

Warfare in Undeveloped and Semi-civilized Countries

106. General

Much of the world is undeveloped, in fact by far the greater part still is, and is likely to remain so for a long time, yet the semi-civilized portions are daily shrinking in size. Even such a wild area as the North West Frontier of India is being influenced by civilization. Roads are opening up, and trade follows the road and in its wake, ghostlike, march luxury hand-in-hand with decadence.

In both these areas—the civilized though undeveloped and the barbaric—military operations fall into two main groups: (1) The maintenance of law and order; and (2) the crushing of revolt. In both of these the predominant factor is space with its correlative time. Where any form of centralized government does not exist, war-like impulses are always contagious. Economic instability, religious fervour and clan customs are all irritants which rapidly generate strife. Therefore, should distances be great, or the time taken in crossing them long, a whole district, or country, may explode into rebellion, or revolt, before even preparations can be made to extinguish the initial outbreak. The problem resembles closely that of a forest fire. The first sparks if tackled in time can be stamped out by one man; if there is no wind the fire will spread gradually, or peter out; but if a gale is blowing no one can say where it will end, unless full preparations have been made to meet such emergencies.

Now, as regards the employment of mechanized and motorized forces in these small war operations, a curious kink exists in the minds of many soldiers. Because the circumstances are primitive, they think that the means of suppressing them must be equally so. I have had it more than once put to me: though tanks may be excellent weapons in Europe, they are likely to prove of little use in Asia. This is illogical. It is true that military organization will have to be modified, but it is not true that in Asia foot or horse solders can traverse greater distances than in Europe, or can march more rapidly. The problem is, as I have already mentioned, one of space and the conquest of space, it is, therefore, pre-eminently a problem for machinery, and the machines which off-hand suggest themselves are the airplane, motor vehicle and scout tank.

As regards the second, its daily radius of action may be reduced at seven to eight times that of the foot soldier, and if scout tanks are mounted on six-wheeled lorries, I see no reason why they should not attain an equal mobility. In May and June 1857, during the Indian Mutiny, the Corps of Guides carried out an extraordinary march of nearly 600 miles in 22 days. With a carefully organized motorized force, I see no reason why this march could not now be carried out in three or four.

As regards transporting the tank on a truck, most soldiers will disagree with me.

[To-day medium tanks, weighing from 18 tons upwards, are, in order to save wear and tear, frequently moved long distances by means of transporters.]

They did so in 1926, when I visited India to examine the North West Frontier from the tank point of view, because, as I believe, they do not realize that the main tank problem in a country like India is not a tactical but an administrative one, hinging as it does on maintenance and repairs.

[This is largely true of all countries and all theatres of war. Unlike a man or a horse, a tank can be "killed" and "resurrected". Until 1942 the German repair organization was incomparably more efficient than our own. The result was that in the North African campaigns of 1941, though General Rommel's tanks were always being "annihilated," he was, nevertheless, generally able to concentrate superior numbers in the field.]

Great distances mean great wear and tear, especially when the climate is hot and the soil sandy or dusty. In such countries each scout tank should be mounted on a truck, and should have three crews of three men carried in a similar vehicle with shelter tents, rations, water, oil, gasoline, ammunition, etc., complete;

each scout tank, two trucks and nine men and their commander forming the smallest tactical unit.

To turn from these details to the first of the two main problems—namely, internal security—the first question is that of organizing the area in question so that disturbances and rebellions may be localized—that is, dealt with immediately by the local police, or if they fail to suppress a rising within twenty-four hours, by military force. With motorized troops this means that, if communications are good, they must be within 150 miles call, and if indifferent from 75 to 100 miles.

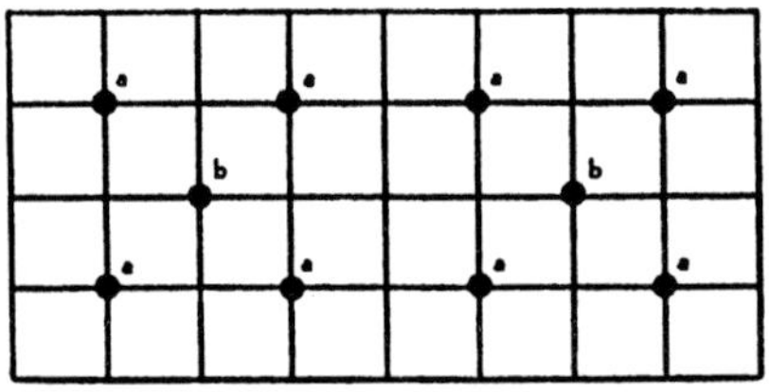

Diagram 20. Organization of Internal Security.

To take a purely formal example: If the country in question were like that to be found in much of the Middle West of America, and the towns and villages were laid out on a grid system, the problem would be a purely geometrical one, as depicted in Diagram 20. Assuming that the area is 800 miles long and 400 broad, and contains 45 towns and cities, one at the corner of each square, then the number of mechanized and motorized forces required will be eight in number—shown on the Diagram by the letters **a.** Only two air centres will be necessary—shown by letters **b.** If each of the letters **a** represent 250 men, and each **b** 200 airmen and 800 soldiers in reserve, then the total for the area of 320,000 square miles will be 4,000. This, of course, is only an explanatory example, but in actual fact in an area of this size I do not think that this figure will be far out.

The reason why so few men are required is due solely to the mobility of their means of transport. In military police work large numbers are seldom required. A few men rapidly brought to the disaffected spot and acting with decision will practically always be able to handle even the most formidable of crowds. In city riots tanks are most useful to surmount barricades, and, if equipped with lachrymatory bomb projectors, will nearly always succeed in clearing houses and centres of resistance, whereas electrified armoured cars have already on more than one occasion proved their value in pushing back hostile mobs.

Generally speaking, in riots and similar turmoils, bullets should not be used, for the object is not to kill but to reinstate police control. Non-lethal gases and smokes should be employed, and never until they have failed to produce the required effects should lethal weapons be resorted to. Had this policy been adopted fifteen years ago in India, few lives would have been lost and riots by now would in all probability have been stamped out. To-day it is as barbarous to shoot down a mob in order to quell a riot as it is to amputate an arm, or leg, without an anaesthetic.

Let us turn to the second problem—namely, crushing revolt, or a war on the frontier. This is a military and not a police problem. Undeveloped countries may be divided into two categories—those which are suitable for motorized forces and those which are not. In the first fall desert warfare and fighting in the plains, and in the second—mountain warfare and forest fighting. In the first two it is obvious that a motorized force will, in nine cases out of ten, prove superior to a non-motorized one. In the second, it is not so obvious that a motorized force will prove of very great assistance, and the reason for this is that most soldiers when they consider the problem of mountain warfare become so obsessed by hill tops that they overlook water springs.

A hill top has certainly a tactical value, but it has little or no administrative value. You can look out and shoot from the top of a hill, but you will seldom find a drink on it, nor will you be able to grow corn on it. The administrative base of the hill top is the valley, and if the valley can be cut off from the hill top, those on it will come down to the valley and surrender. Hitherto, to sit in the valley has been an unpleasant task, because the man on the hill top could shoot into it with effect. Deny him this power and the hill top becomes a trap. This, to a large degree, can be accomplished by the use of bullet-proof armour. Mountain warfare is really a misnomer, its correct name is Valley warfare.

107. Mountain Warfare

Mountain warfare proper—namely, clambering up hill sides and engaging the enemy on their slopes and summits—is obviously a type of fighting totally unsuitable for motorized and mechanized troops; but valley fighting falls into another category. To my knowledge, on the North West Frontier of India the majority of valleys and nullahs can be negotiated by tanks, and many of them by motor vehicles; therefore, as most of the villages are situated in them, they can be attacked by tanks.

In Lecture XIV of my "Lectures on F. S. R. II" I have dealt in sufficient detail with the tactics of tanks in nullah and village fighting to make it unnecessary to refer to them again. Instead, I will make a suggestion as regards air tactics.

[See Appendix II.]

Hitherto the airplane has prove itself to be the most important instrument of reconnaissance in this type of warfare, but as a lethal argument it has not proved itself a great success, and curiously enough because it uses lethal weapons—high exposive bombs. To hit a small village covering an acre or two from a bullet-safe altitude is not an easy thing to do, and if the village is hit, it is unlikely that more than three or four men, twice as many women and children and a few goats will be killed and wounded; for if war is in progress most of the fighting men will certainly not be there. Instead, what should be attempted is that every bomb dropped should fall close enough to the village to "hit" every man, woman, child and goat in it without doing permanent injury to any of these creatures.

Ever since the World War we have possessed a toxic smoke which will cause acute toothache lasting for a number of hours. Given this chemical and not high explosive, here is the picture:

War is declared and the warriors bid farewell to their wives and children. They swarm up the hill sides and lie in wait for the tail end of the punitive column. No column appears, instead a solitary bombing machine loaded with half a ton of toothache mixture sails over their village, noses for wind and drops its goods on the windward side. Growing hungry the warriors return for their evening meal; but no curried goat and chupatties await them, instead a howling mob of women and children who hurl expletives at their heads, words which have more effect on their morale than ever had bullet or T.N.T. After six such vocal bombardments the war is off. Three tons of toxic smoke and a little gasoline, costing perhaps a hundred pounds, has accomplished in a few days what a two million pounds punitive column has frequently failed to accomplish in three months.

Why do we not adopt such a common-sense method? Because we lack imagination, and because we do not possess the courage to face public opinion and do it.

BUSH FIGHTING

108. Conditions Affecting Fighting in Bush and Forest Countries

Bush warfare, in contradistinction to warfare in scrub is not applicable to motorized troops, its limiting factor being the spacing of the trees and the thickness of the undergrowth. As I have already mentioned, scout tanks should be able to zig-zag their way through most forests, but that normally they will be able to do so rapidly is unlikely; therefore, in these operations they can work in close cooperation with light infantry without sacrificing their speed.

Forest operations fall into two categories: (1) Where the whole theater of war is covered with trees, such as in north Russia, and (2) where it is broken up by a belt of forest land such as the Wilderness in Grant's Virginia campaign of 1864, or by large isolated patches of forest, as at the battle of Königgratz in 1866, or the fighting in the Ardennes in 1914.

In the first case, as the forest cannot be avoided, light infantry should work forward of the tanks, these machines being held in support. If the light infantry are held up, the tanks should move forward, and once the obstruction is cleared, they should fall back into support. Diagrams of certain formations which can be used are given on page 154-155 of my "Lectures on F. S. R. II."

[See Appendix III.]

In the second case the patches of forest may be avoidable—that is to say, a way round them may be discovered. Should this be the case, it will frequently be more profitable to hold their exits and move on, reducing their garrisons by starvation, air bombardment, or gas attack, rather than by attempting to clear them out, which as often as not is likely to be a slow operation. Never waste time, in war time in even more precious than human lives.

109. Convoy Camps

In undeveloped lands the main problem is always that of supply. The normal duty of the troops is escort work and not fighting; therefore the protection of the rear services, supply column, etc., both on the line of march and at rest is of first concern.

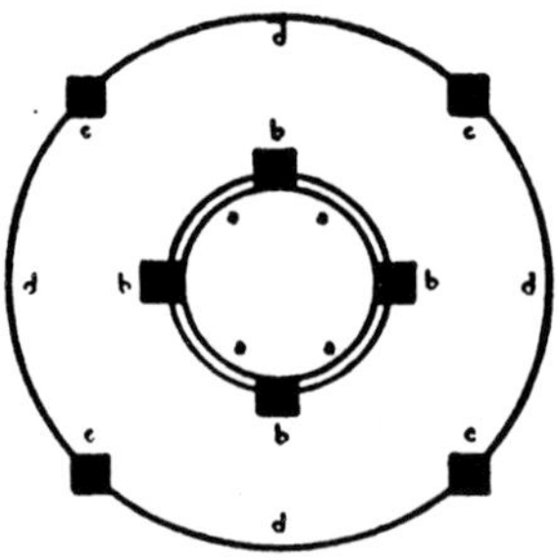

Diagram 21. Motorized Laager.

When a halt takes place, the normal method is to construct a defended camp, with the non-combatants in its center and the combatants on its perimeter, or in reserve ready to sally forth and attack an enemy who has broken through the line of picket posts. When a motorized and mechanized force is used the principles of defenses will be the same; but the method of applying them will be easier. First, the supply trucks should be parked in a circle, or square, with their engines pointing inwards so as to protect them from bullets, and within this laager the non-combatants should camp. The outer ring of defenses should be formed by detachments of armored cars with scout tanks patrolling in between, a reserve of these machines being kept within the laager.

A very simple example of such a laager is given in Diagram 21. The lorry laager is shown by the letters **a** immediately protected by four flanking bastions—**b.** Outside this runs circle **d,** the area to be patrolled by the scout tanks, with its four armored car bastions marked by the letters **c.**

Generally speaking, it will be advantageous to form the laager in a dip in the ground, so that it is hidden from direct observation. If the enemy has tanks, natural tank obstacles should be looked for, and if he has aircraft, the air-defense troops should be located at **c.**

DESERT WARFARE

110. General Principles

In desert warfare, as the two predominating factors are water and the enemy's speed of movement, it is obvious that motorization should prove of the highest value; for water can be carried in bulk, and the motor vehicle, armored car and scout tank should generally be able to out-distance any horseman.

[This has proved true.]

Bearing these two advantages in mind, and combining them with aircraft reconnaissance, desert warfare in the future should become a very simple operation.

The tactics I suggest are as follows: Two forces are required —a tactical dust-pan and a broom.

[General Rommel's attack in May-June, 1942, was of this type. His plan was as follows:

(1) To engage his enemy's right at Gazala and slowly push it back with his left wing (pan).

(2) To sweep round his enemy's left at Bir Hakeim with his own right wing (broom): cut the British communications running eastwards, and then, as his enemy could not escape to the north, sweep him westwards into the dust-pan.]

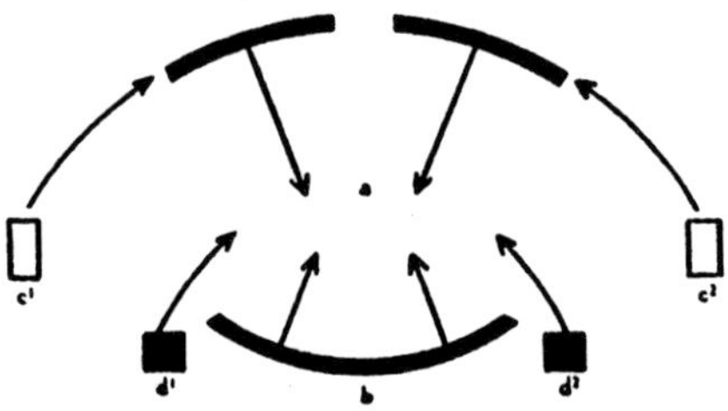

Diagram 22. Desert Warfare Tactics.

The first, consisting of motorized light infantry supported by a group of armored cars on each of its wings; the second, of two columns of armored cars, scout tanks and motorized light infantry, each forward and well out on the two flanks of the dust-pan. Look at Diagram 22. By aircraft the enemy is reported to be at **a**; at once the dust-pan **b** deploys and **slowly** moves in his direction. Meanwhile the two brooms (**c**¹ and **c**²) move forward out of sight of his flanks towards his rear, and deploy in a semi-circular formation; then they turn inwards and sweep the enemy towards the pan. Should the enemy attempt to break out on the flanks of **b**, then the two armored car groups **d**¹ and **d**² will either block these flanks or drive him back into the enclosing circle. These tactics are but a modification of those used by Alexander the Great against the Scythians as described in my "Lectures on F. S. R. II."

[See Appendix IV.]

LECTURE XV

CHAPTER XI

Movements by Sea, Land and Air

As these last three chapters of "F. S. R. II" deal with routine work, I will only examine them briefly, not that they are unimportant, but because their importance is so great that they might well constitute a separate volume of their own. The first point to note in Chapter XI is that during the last thirty years movement has completely changed. Not only was air power then unknown, but during this single generation the whole aspect of civilization has been transformed by the motor vehicle. It has been changed so completely that, mentally not being able to keep pace with it, we do not realize it, and it is because of this want of appreciation that at gatherings like the Geneva Disarmament Conference, such unlimited clap-trap is talked about the newer arms. To abolish tanks and fighting aircraft simply means that agricultural tractors, motor vehicles and civil aircraft will replace them. A Fordson tractor can be converted into an indifferent tank in a few days; a truck can be converted into an armored car, and a civil airplane into an indifferent bombing machine. All these indifferences are but relative; so long as tanks, armored cars and fighting aircraft exist, indifference exists; but abolish these newer weapons and indifference vanishes, and is at once transformed into efficiency. The truth is that all these hopeless suggestions are made by men with either their tongue in their cheek, or by men who cannot see that, throughout history, the tools of peace have **always** been converted into the weapon of war. As the billhook was father of the medieval billmen, the hunting bow of the long bow, the sporting rifle of the military rifle, so is the civil airplane the father of the bombing machine, the touring car of the armored car, and the field tractor of the tank. If we are deprived of these sons, the fathers will fight instead, and the only difference will be that war will be old fashioned instead of new fashioned.

No, mechanical movement at sea, on land and in the air is with us, and will remain with us so long as it proves useful, not to the soldier **but to the civilian.**

At sea, movement has remained much the same during the last fifty years, and when war was declared special vessels were not required for the transportation of armies. In mechanized warfare this condition will continue, but with one exception. If landings are to be made rapidly with amphibian tanks, these machines should be embarked upon specially constructed tank-carrying ships, built on somewhat similar lines to the large steam whaling vessels.

[Special landing craft have now been built to carry non-amphibian tanks and enable them to land on a beach. They bear no resemblance to the vessel here suggested. The heart of the landing problem is to make them swim.]

Instead of hauling whales in and out of the ships, tanks under their own power will crawl out and into them. The ship will in fact be nothing more than a mobile tank-roost.

With the larger types of tank the main difficulty is getting them on to and off the deck of the ship's hold. Once there, they are the easiest handled of cargoes, for they can stow themselves away and un-stow themselves when time comes for disembarkation.

Land movements may now be classed under three definite headings—namely, movements by road, rail and across country. Roads will generally be allotted to the motorized columns, the mechanized arms moving alongside them if necessary. Motorized units should, therefore, be related to trucks, by which I mean that a small unit, like a platoon, should not consist of an arbitrary number of men, or even a tactical number, but of a logistical number—that is, it should represent a truck load. If a truck holds 20 men, then the strength of the platoon should be 20 men. Units thus organized will be easier to move, and further still, organization will be maintained during movement.

By rail, tanks are as easily transported as by ship, though the heavier machines will require specially built trucks to carry them.

They may either be moved on trucks or slung between two sets of bogies. To get them on to a train, all that is necessary is a stoutly built entraining ramp constructed of steel girders or wooden sleepers. A well drilled tank company should be able to entrain sixteen machines in from five to ten minutes.

Cross-country movements by land trains I have already touched upon in brief. Their importance, I think, is a growing one. As I have pointed out, the land train of the future ought to be able to haul upwards of 100 tons at the average speed of eight miles the hour, or 100 miles in about twelve hours. It is not necessary for me here to enter into the depot and convoy schemes, as I have already fully described them.

Lastly, to turn to the air, here we find movement daily growing in importance. A few years ago, the picture of several hundreds of people being evacuated from Kabul by air was a truly astonishing one, and foreshadowed enormous military possibilities.

[These possibilities have been more than verified, **e.g.**—the air invasions of Norway, Crete, etc.]

It is not too much to suppose that considerable numbers of soldiers, supplies in bulk, and even scout tanks, will in the future be transported from place to place by air, and that in undeveloped countries aircraft will be considered the normal means of evacuating the wounded. Not even the most gelatinous of humanitarians could object to a flying hospital of 100 beds. But forbid the use of fighting aircraft, then with no great difficulty could such a machine be converted into a thirty-ton bomber. War is a strange business, but solving peace problems seems to be a stranger one.

CHAPTER XII

Orders, Instructions, Reports and Messages

As I have pointed out more than once, orders, instructions, reports and messages will have to abandon their many official frills and step out stark naked into the reality of war. The object of an operation order is to impart information you cannot actually convey by voice. It may be the word "move," or "halt," or it may be a long rigmarole; in either case it is seldom necessary to turn it into a ritual so holy that it is considered almost sacrilegious not to begin an operation order with "information" . . . "intention," and go on, etc., etc.

All orders will have to be as brief as possible, and not as formal as possible. They should be based on a profound appreciation of possibilities and probabilities, which, as I have explained, will generally lead to a series of alternatives. Therefore an order should not be suited to one operation but to several possible phases of this operation. It should possess a central idea and several radii working out towards the final circumference—victory to you and defeat to the other man.

If we wish to prepare ourselves for mechanized warfare, it is time we broke away from existing conventions, substituting common-sense for ritual. A methodical soldier may be able to find everything, like a tidy person. This is excellent, but what is infinitely better is being able to make use of things instantaneously—anything, ground, tanks, infantry, broomsticks. What above all the fighting soldier requires is not a brain which works by rules, but a brain which rules by work—that is, immediate action.

A great deal of this training in spontaneity of action will depend on our orders and instructions. In the future much more must be left to the initiative of the individual than in the past.

[The various North African campaigns have, again and again, accentuated the wisdom of this.]

ARMORED WARFARE

Though the central idea must be maintained, actions should be as flexible as possible. Reports must be as brief as possible and should always, when possible, suggest actions. To state that the enemy is blowing his nose may be interesting, but to report that he is looking eastwards and is open to a backside kick from the west is some thing of real importance. Messages should be in code, and when sent between units in battle they should generally be in clear.

[This was proved correct in the rapidly moving North African battles.]

Time, time, and the saving of it, should be the soul of every order and instruction, of every report and of every message.

CHAPTER XIII

Intercommunications

The last problem dealt with in "F. S. R. II" is that of intercommunication, and it forms the first which will have to be threshed out before "F. S. R. III" can be written. Without a thoroughly efficient system of intercommunication there can be no proper control, and without control every increase in speed can but make confusion worse confounded.

I am unable here to offer any definite suggestions, for though it is easy enough to work out systems of intercommunication, it is not possible, without a long period of test and trial, to discover which system is best, and when a certain system is applicable.

The whole problem may be examined under four headings:

(1) Intercommunication between the members of a tank crew;
(2) Between individual tanks;
(3) Between tank units;
(4) And between field and rear headquarters.

As regards the first, this has been solved by an instrument called the laryngaphone; as regards the second, colored flags and semaphore signalling have been tried out with considerable success; also for night communication colored lights were used during the war, but as yet neither the semaphore nor colored lights have been fully tested out.

As regards the last two, wireless would appear to be the only satisfactory solution. This means two things: The first, that each group of tanks must be controlled by a command tank equipped with a sending and receiving set; and the second that the tanks of each group must operate as a unit, a single machine never being long out of sight of the unit commander's.

[This now applies to all British tanks, and not only to the command tanks.]

The present difficulties in intercommunication are due more to system than to apparatus. When a definite tank tactic is elaborated and cut clear from the present idea of cooperating with infantry; when the assault is no longer the goal of every attack; when initiative is looked upon with as high favor as obedience; when leaders learn to think more rapidly and express their ideas in fewer words, and when tank maps are issued, then and then only shall we be able to put the apparatus we have to a true test—a test which will prove out their value for mechanized operations, instead of for slow moving infantry warfare. The first problem here, as everywhere, is to cultivate a new angle of thought, to divorce our minds from the stage-coach epoch of warfare, so that we may gaze into the future with a clear and unprejudiced mind. With this supreme military requirement as my last words, I will end this series of lectures on a book which I hope before long we may all see.

APPENDICES

APPENDIX I

The Theory of Penetration

"The fundamental difficulty in an attack of penetration is continuity of advance, and this is restricted more through hostile flank pressure than hostile frontal resistance; for it is an interesting fact . . . that in the last war the flanking angles of the salient formed by an attempted penetration were normally 45 degrees, and never more. What does this mean? It means that the problem of penetration must be founded on a mathematical calculation involving the factors of the depth to be penetrated and the nature of the enemy's defenses. These—bearing in mind the sloping inwards of the flanks of the attack at 45 degrees—will enable us to decide whether we have sufficient men for the required frontage of attack . . . Thus, in

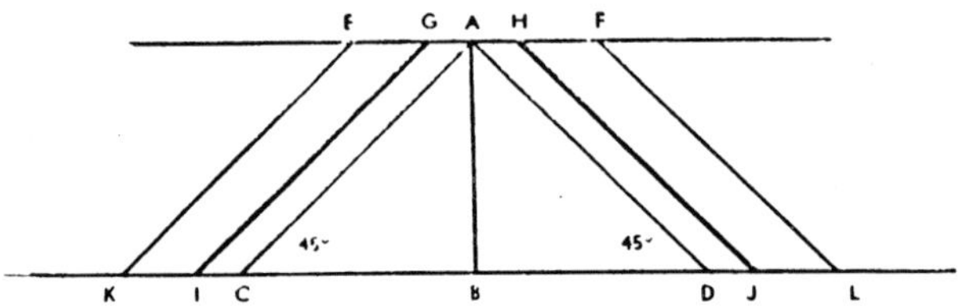

Diagram 23. Theory of Single Penetration.

Diagram 23, if A-B—the depth of the enemy's defenses—is five miles, then we know that the front C D to be attacked is ten miles. But to push a few men through at A is not sufficient, for the gap created must be wide enough to enable the exploiting forces to be moved rapidly forward, which means outside of hostile machine gunfire. Such a gap must be at least five miles wide—E F. This will allow of a frontage two miles wide—namely, G H—fairly free from projectiles, as well as a considerable area, G I J H, also free to assemble troops in. Therefore we must add two and a half miles to each flank of the original ten miles frontage of attack extending C D to K L; K L is, therefore, fifteen miles.

"Having worked this problem out, we must examine the enemy's defenses, divide them into objectives, allot an attacking force to each, and decide upon what force we require for exploitation. Lastly, we must total up these figures and see whether we have sufficient men. If we have not, then to attack is to court disaster. During the war scores of thousands of men's lives were thrown away because it was not realized that the flanks of an attack of penetration slope inwards at angles of 45 degrees.

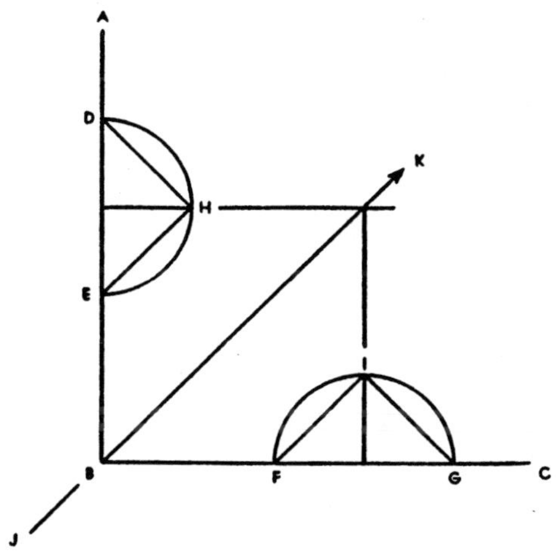

Diagram 24. Theory of Dual Penetration.

"Not possessing the required number of troops, what is the solution? There are two—namely, the dual penetration and the tank penetration. I will examine these in turn.

"Granted that somewhere on the enemy's flank there is to be found a salient—A B C in Diagram 24. Then, should two frontal attacks be launched from D E and F G, not with the idea of penetrating his defenses on these fronts, but instead of rendering the area H E B F I untenable, this area will have to be evacuated, and by rapidly following up this evacuation a penetration will be effected along the line J K.

"The difficulty in this method of attack is once again the large number of men required, because both attacks have to be made in strength in order to draw in large forces of the enemy, so that he is unable to oppose the advance along the line J K.

"This method is in fact a makeshift, for it does not overcome the main difficulty, which is the sloping inwards of the flanks. If instead we could make these flanks slope outwards, how easy the problem would be. This is exactly what the tank enables us to do, because it is bullet-proof. . . .

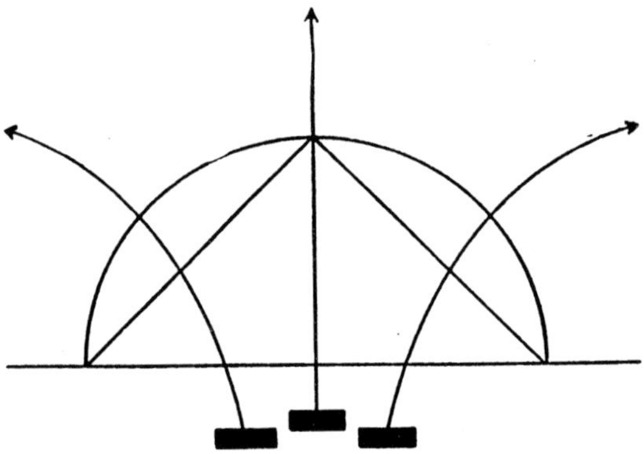

Diagram 25. Theory of Tank Penetration.

". . . The tanks move off in three bodies, see Diagram 25. The central one goes straight forward . . . the two flanking ones enter the dent and then wheel outwards and roll up the rear of the enemy on each side of it. What does this lead to? The automatic widening of the initial front instead of the automatic dwindling of the advancing front as the penetration grows deeper and deeper.

"In the dual penetration this ability to push out the flanks reaches its maximum effect, because it enables the two attacks to be separated by a greater distance. The area threatened between the two attacks is also greater; therefore the wound made in the enemy's defenses is more extensive, so extensive that he may well bleed to death."

APPENDIX II

Tank Nullah Tactics

"Turning now to an attack on a defile, what are the main difficulties? The defile cannot be avoided, because the nullah running through it is the line of supply. The enemy occupy its two flanking heights. To attack him frontally means heavy

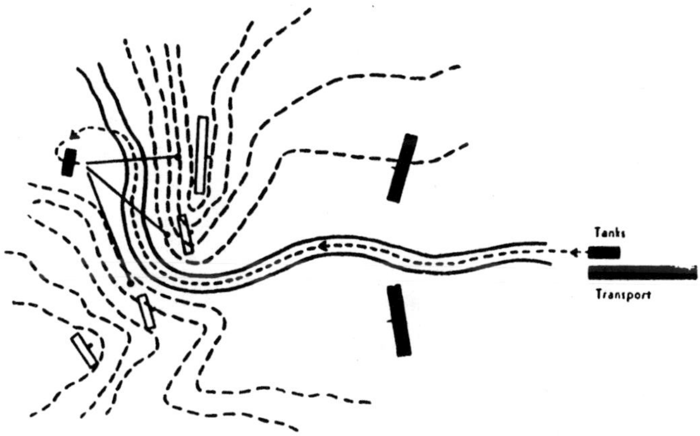

Diagram 26. Tank Nullah Tactics.

loss; to outflank him is often impossible; yet should he be outflanked—that is, threatened in rear—in nine cases out of ten he will abandon his position. I examined many of these defiles on the North West Frontier of India in 1926, and nearly all presented little or no difficulty to the advance of light tanks, which in a few minutes could have moved up the nullah and by mortar and machine gun fire have taken the enemy in rear (see Diagram 26). The value of such an attack is that the enemy will be caught between two fires and his line of retreat threatened if not cut. I feel certain myself that once hillmen have been attacked in this way, on the next occasion, directly they see the tanks advancing they will bolt."

APPENDIX III

Bush and Forest Warfare

"In this type of warfare all columns should be kept well closed-up. The advanced guard should be close to the head of the main body, and the rear guard should be strong, because savages are prone to attack a column in rear. The advanced guard and rear guard should be organized in three groups (see Diagram 27), the advanced group and two flanking groups.

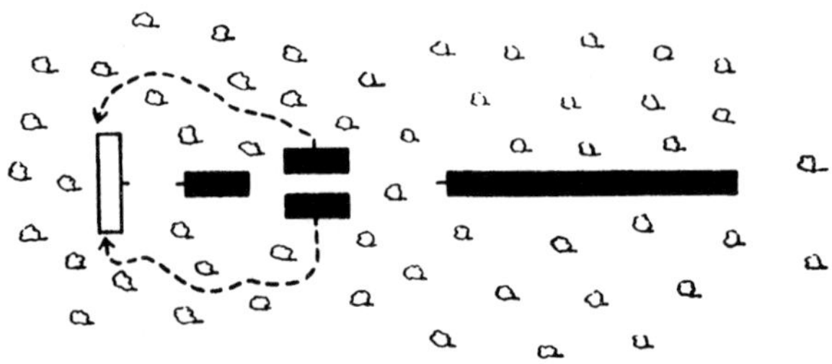

Diagram 27. Bush Advanced Guard Tactics.

Directly the advanced group, or point, is fired upon, it should halt and hold the enemy with its fire, while the flanking groups move outwards, and, under cover of this fire, attempt to turn the enemy's flanks and take him in rear. This operation is once again but a special reading of our old friend—hold, maneuver, hit.

"In forest fighting the soundest formation to adopt is that of the diamond, or elastic square; an advanced guard, two flank guards and a rear guard, each group prepared to assume any one of these three duties. Thus, in Diagram 28, in a column advancing with B as advanced guard, if A is attacked, this

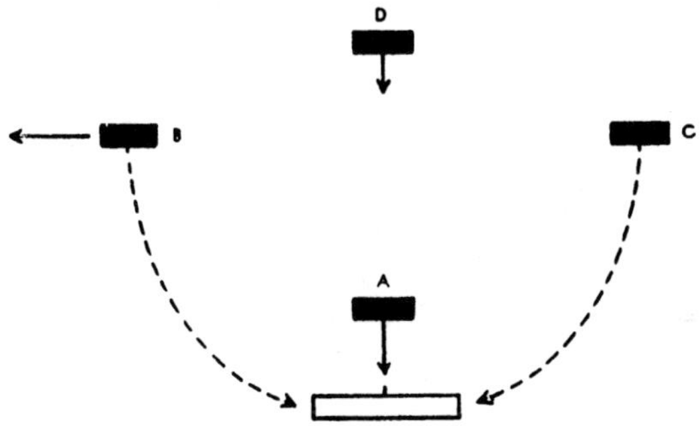

Diagram 28. Forest Warfare Formation.

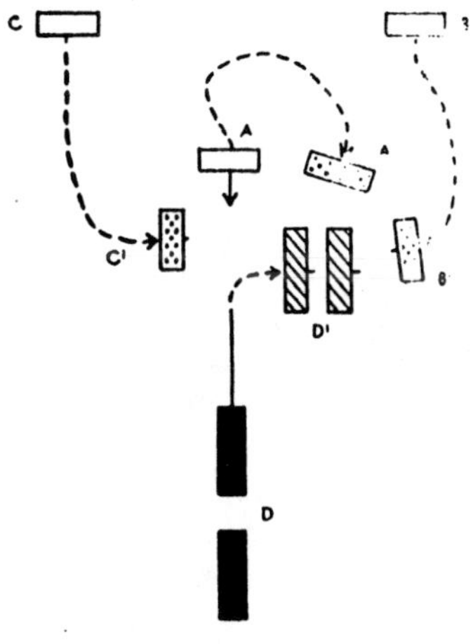

Diagram 29. Battle of King's Mountain, 1780.

group holds, and B and C maneuver towards the enemy's flanks, while D remains in reserve. If B is attacked, then this group holds, and A and D maneuver, and so on.

"An interesting example of these tactics occurred in 1780 at King's Mountain during the War of the American Rebellion. Major Patrick Ferguson at the head of several hundred English troops and local militiamen was confronted by an equal number of American backwoodsmen under Morgan, who organized his force into three groups—A, B and C, see Diagram 29. When A came into contact with Ferguson's column D, it fell back, when, as D advanced against it, D was suddenly threatened in flank by B, and changing front to meet this threat, A turned about and struck at D's left flank, whereupon C fell upon D's rear. A prettier piece of forest tactics it would be hard to devise; further, it is as applicable today as it was then."

APPENDIX IV

Alexander the Great's Anti-Scythian Tactics

"The controlling factor in desert warfare is scarcity of water, therefore desert tribes, who are nearly always mounted, do not usually attempt to assault or to push home an attack, but, instead, by circling round and round their enemy, they aim at slowing down his advance between two watering points, in order to compel his capitulation by offering him no alternative to surrender except death from thirst.

"This method of warfare is as old as war itself, and again and again have the Scythians, Parthians and Arabs annihilated great armies by means of it, as Crassus was annihilated at Carrhae in B.C. 53. Alexander the Great, who was never at a loss for a tactical solution, whether confronted by an organized army, guerillas, hill tribes, or a siege, defeated the mosquito attack of the Scythian by a tactic which for cunning has seldom been equalled. First he threw out, well ahead of his main body, a strong force of light cavalry and horse archers—A in Diagram 30. Seeing these men, at once the Scythians broke their horde formation and began circling around them. Next Alexander advanced his infantry in half-moon formation—B, and drew up his heavy cavalry—C behind them. Suddenly, at a signal, his heavy cavalry galloped forward through and around the infantry, and

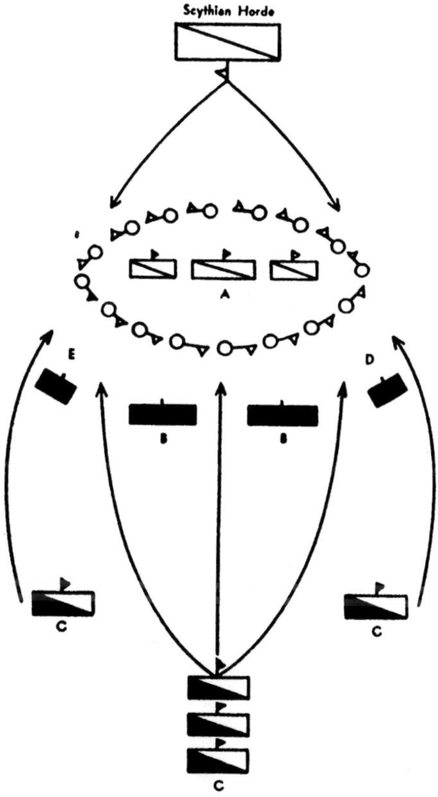

Diagram 30. Alexander's Anti-Scythian Tactics.

filling up the gaps D and E bottled up the Scythian horse between the light cavalry and the infantry. This done, the enemy was speedily slaughtered."